BIOLOGICAL
EFFECTS OF
LOW
LEVEL
EXPOSURES:
DOSE-RESPONSE RELATIONSHIPS

Edward J. Calabrese
Editor

CRC Press
Taylor & Francis Group
Boca Raton London New York

CRC Press is an imprint of the
Taylor & Francis Group, an **informa** business

CRC Press
Taylor & Francis Group
6000 Broken Sound Parkway NW, Suite 300
Boca Raton, FL 33487-2742

© 1994 by Taylor & Francis Group, LLC
CRC Press is an imprint of Taylor & Francis Group, an Informa business

No claim to original U.S. Government works

Visit the Taylor & Francis Web site at
http://www.taylorandfrancis.com

and the CRC Press Web site at
http://www.crcpress.com

Edward J. Calabrese is a board certified toxicologist who is professor of toxicology at the University of Massachusetts School of Public Health, Amherst. Dr. Calabrese has researched extensively in the area of host factors affecting susceptibility to pollutants, and is the author of more than 270 papers in scholarly journals, as well as 24 books, including *Principles of Animal Extrapolation; Nutrition and Environmental Health*, Vols. I and II; *Ecogenetics; Safe Drinking Water Act: Amendments, Regulations and Standards; Soils Contaminated by Petroleum: Environmental & Public Health Effects; Petroleum Contaminated Soils*, Vols. 1, 2, and 3; *Ozone Risk Communication and Management; Hydrocarbon Contaminated Soils*, Vols. I, II, and III; *Hydrocarbon Contaminated Soils and Groundwater*, Vols. 1, 2, and 3; *Multiple Chemical Interactions; Air Toxics and Risk Assessment; Alcohol Interactions with Drugs and Chemicals; Regulating Drinking Water Quality; Biological Effects of Low Level Exposures to Chemicals and Radiation; Contaminated Soils: Diesel Fuel Contamination; Risk Assessment and Environmental Fate Methodologies; Principles and Practices for Petroleum Contaminated Soils;* and *Performing Ecological Risk Assessments.* He has been a member of the U.S. National Academy of Sciences and NATO Countries Safe Drinking Water committees, and of the Board of Scientific Counselors for the Agency for Toxic Substances and Disease Registry (ATSDR). Dr. Calabrese also serves as Chairman of the International Society of Regulatory Toxicology and Pharmacology's Council for Health and Environmental Safety of Soils (CHESS) and as Director of the Northeast Regional Environmental Public Health Center at the University of Massachusetts.

Preface

The biological effects of low level exposures to chemicals and radiation has been an ever-increasing focus of research and regulatory attention in the areas of environmental and occupational toxicology and epidemiology, radiation biology, and pharmacology. While controversy has long centered on the challenges of confidently extrapolating from high to low dosages, it is the belief of the BELLE Advisory Committee that this issue must be directly addressed via rigorous research methodologies in the low dosage areas. Such an experimental focus is not only providing insights of relevance to regulatory agencies, but also to those basic research scientists interested in molecular mechanisms involved with cellular adaptations.

The present book, *Biological Effects of Low Level Exposures: Dose-Response Relationships*, represents the proceedings of a conference held on April 26–27, 1993 in Crystal City, Virginia. The proceedings are organized in four parts: Part I, Historical, Philosophical and Conceptual Foundations of BELLE research; Parts II and III, Biological Effects of Low Level Exposure to Chemicals (Part II) and Radiation (Part III); Part IV provides conference perspectives and summary.

BELLE Advisory Committee Members
August 1993

Edward J. Calabrese, Ph.D.
Chair, BELLE Advisory Committee
School of Public Health—N344
University of Massachusetts
Amherst, MA 01003

Mary F. Argus, Ph.D.
U.S. EPA, TS-796
401 M Street, SW
Washington, DC 20460

James Robert Beall, Ph.D.
ER-72 GTN, MS-G236
Office of Health and
    Environmental Research
U.S. Department of Energy
Washington, DC 20585

Ralph Cook, M.D.
Dow Corning Corporation
Mail #CO1120
Midland, MI 48686-0994

C. Richard Cothern
Center for Environmental
    Statistics Development Staff
U.S. Environmental Protection
    Agency, PM-223
Washington, DC 20460

J. Michael Davis, Ph.D.
Health Scientist
U.S. EPA, MD-52
Environmental Criteria and
    Assessment Office
Research Triangle Park, NC 27711

Max Eisenberg, Ph.D.
Executive Director
Center for Indoor Air Research
1099 Winterson Road, Suite 280
Linthicum, MD 21090

William Farland, Ph.D.
U.S. EPA, RD-689
Office of Health and
    Environmental Assessment
401 M Street, SW
Washington, DC 20460

James R. Fouts, Ph.D.
National Institute of
    Environmental Health Science
P.O. Box 12233
Research Triangle Park, NC 27709

John Frawley, Ph.D.
President
Health and Environment
    International, Ltd.
400 West 9th Street, Suite #401
Wilmington, DE 19801

Hank S. Gardner, MSPH
Research Biologist
Health Effects Research Division
U.S. Army Medical Bioengineering
    Research & Development Lab
Fort Detrick
Frederick, MD 21701-5010

Ron W. Hart, Ph.D.
Department of Health and
    Human Services
Director, National Center
    for Toxicological Research
NCTR Drive
Jefferson, AR 72079

A. Wallace Hayes, Ph.D.
Gillette Company
Prudential Tower Building-45th Floor
Boston, MA 02199

Donald H. Hughes, Ph.D.
Procter & Gamble Company
Miami Valley Laboratories
P.O. Box 398707
Cincinnati, OH 45239-8707

John G. Keller, Ph.D.
P.O. Box 768
Olney, MD 20830-0768

Dan Krewski, Ph.D.
Health & Welfare Canada
Environmental Health Center
Room 109
Tunney's Pasture
Ottawa, Ontario K1AOL2
CANADA

Roger O. McClellan, D.V.M.
President
Chemical Industry Institute
  of Toxicology
P.O. Box 12137
6 Davis Drive
Research Triangle Park, NC 27709

Leonard Sagan, M.D.
Electric Power Research Institute
3412 Hillview Avenue
Palo Alto, CA 94303

Harry Salem, Ph.D.
Toxicology Division, CRDEC
U.S. Army, SMCCR-RST
Aberdeen Proving Grounds, MD 21010-5423

Andrew Sivak, Ph.D., A.T.S.
Environmental Health Sciences
P.O. Box 1038
Kendall Square
Cambridge, MA 02142

Lester Smith, Ph.D.
Agency for Toxic Substance and
  Disease Registry-Mail Stop E-29
Division of Toxicology
1600 Clifton Road, NE
Atlanta, GA 30333

Richard Dean Thomas, Ph.D.
National Research Council, Tox/Epi
National Academy of Sciences
2101 Constitution Avenue—JH653
Washington, DC 20418

D.W. Whillans, Ph.D.
Ontario Hydro
Health and Safety Division
1549 Victoria Street, East
Whitby, Ontario L1N 9E3
CANADA

Contents

PART IV: BELLE CONFERENCE PERSPECTIVES AND SUMMARY

BIOLOGICAL EFFECTS OF LOW LEVEL EXPOSURES:

DOSE-RESPONSE RELATIONSHIPS

PART I

Historical, Philosophical, Conceptual Foundations of BELLE Research

PART I

Historical, Philosophical, Conceptual Foundations of BELLE Research

CHAPTER 1

Changing Scientific Paradigms

Kenneth F. Schaffner, Philosophy Department,
George Washington University, Washington, DC

INTRODUCTION

This chapter contains discussion of scientific change, focusing on the themes developed in Thomas S. Kuhn's influential book *The Structure of Scientific Revolutions*.[1] I will sketch a brief history of the reactions to Kuhn's work, outline the main elements of his theory of scientific revolutions, and comment on some of Kuhn's most recent thoughts as expressed in his 1992 Rothschild lecture.[2] The essay considers at several points the applicability of Kuhn's classical and most recent views on the controversy over the "present linear paradigm" versus hormesis and U-shaped approaches to toxic agents. I conclude with a brief discussion of some important methodological differences between the philosophy of the physical sciences — the source of Kuhn's examples — and the biological sciences, and relate these differences to the themes of this BELLE conference.

Kuhn's book *The Structure of Scientific Revolutions* first appeared in 1962, and in a second edition with an important Postscript in 1970. The book has attracted enormous attention and strong admiration and criticism, sometimes from the same individual. For example, the philosopher of science Ronald Giere writes that when he first encountered Kuhn's book in 1962 he found it "philosophically unacceptable."[3] He adds later in his 1988 book, however, that "with a quarter century of hindsight it is now clear that Kuhn's *Structure of Scientific Revolutions* . . . is, by any measure, the most influential book on the nature of science yet to be published in the twentieth century." This view is not limited to the philosophical community. Many scientists have found Kuhn's analysis illuminating and Kuhn's term for a notable scientific theory — *paradigm* — has become a common word in specialist scientific journal articles. Even social reformers and politicians have accepted Kuhn's views with enthusiasm. During the mid-'60s the "Free Speech Movement" at the University of California at Berkeley relied on Kuhn's book to legitimize their opposition to and distrust of the older generation and the status quo. More recently, the *Washington Post* reported that *The Structure of Scientific Revolutions* was Vice President Al Gore's "favorite book."[4]

In the following section I will outline what I take to be the basic themes of Kuhn's classical view in an attempt to indicate why this philosophical analysis has had such influence. I will then discuss more recent developments of Kuhn's thought, including his concerns about misinterpretations of his account.

KUHN'S CLASSICAL VIEW OF SCIENCE AND SCIENTIFIC REVOLUTIONS

One of the major debates over the past nearly 30 years in the philosophy of science has been the extent to which scientific change is rational and represents objective progress. This debate was largely initiated by the contributions of Kuhn and Paul Feyerabend.[5] Feyerabend appears to have developed his ideas from his association with, and initial commitment to, Sir Karl Popper's philosophy of science, and also from his acquaintance with the arguments pro and con regarding Bohr's "Copenhagen" interpretation of quantum theory.[6] Kuhn's prior work was in physics and the history of science, and while he was a junior fellow at Harvard he produced a clear and eminently readable account of the Copernican revolution.[7] During the late 1950s and early 1960s, Kuhn and Feyerabend were both associated with the philosophy department at the University of California at Berkeley, and through extensive conversations developed their initially fairly similar views of the relation of scientific theories and of scientific revolutions. Important aspects of their views had been anticipated in the monographs of N. R. Hanson[8] and S. Toulmin.[9]

Though Feyerabend's philosophy has had its own significant influence, and though both Feyerabend and Kuhn share some important common views, it was Kuhn's new conception of a *paradigm* and his association of a *shift of paradigms* with dramatic discontinuity and irrationality in science that have most captured the imagination of nonphilosophers, and in the essay that follows, my major focus will be on Kuhn's work. I will begin by attempting to provide an answer to the question, "What is a Kuhnian paradigm?"

The Elements of a Kuhnian Paradigm

In the 1962 version of his classic, *The Structure of Scientific Revolutions*, Kuhn introduced the notion of a "paradigm" that was both prior to and more general than a scientific theory. In defining this idea of a paradigm, Kuhn differentiated this notion from the more traditional concept of a scientific theory. In his book, *The Structure of Scientific Revolutions*, Kuhn made the then startling suggestion that the *standards* by which groups of scientists judge theories are heavily conditioned by the theory itself, or, in Kuhn's term, by the *paradigm*. A paradigm is much broader than a theory — it contains within itself not only the abstract or formal theory per se, but also metaphysical commitments — assertions as to what things really exist — as well as a methodology, specific examples of application, and standards of evaluation. Thus an Aristotelian would find *unintelligible* basic Newtonian notions of vacuums and force-free motions. Similar instances in the history of science can also be found in Einstein's opposition to the essentially statistical character of quantum mechanics, in the furor over Darwin's evolutionary theory in England and America in the 1860s and 1870s (see Hull[10]), and in the debate in the 1960s between the instructive and the clonal selection theories of immunology (see Schaffner[11]). Current debates about hormesis and the role of U-shaped or J-shaped dose-response curves may also illustrate these themes. I shall say more about this below, but addressing the major features of that debate will be the task of others in these chapters.

In addition to containing these elements of formal theory, metaphysics, and standards, Kuhn also characterized a paradigm as a "concrete scientific achievement" that functioned as a "locus of professional commitment" and which was *"prior* to the various laws, theories, and points of view that . . . [could] be abstracted from it" (my emphasis). Such a paradigm was usually embodied in a scientific textbook or, earlier, in books such as Ptolemy's *Almagest* or Newton's *Principia*, but obviously it could also be found in a (collection of) scientific article(s). This concrete scientific achievement aspect of a paradigm has at least two significant but interacting components; one is sociological and the other is temporal.

In introducing the notions of professional commitment and scientific communities as defined by (and defining) a shared paradigm, Kuhn identified an important characteristic of scientific belief systems that has had a major impact on the sociology of science. In stressing a temporal developmental aspect of a paradigm — normal science is the solving of a variety of puzzles not initially solved by the paradigm, as well as the paradigm's further elaboration and filling in of fine-structured details — Kuhn focused attention on the temporally extended and subtly varying features of a scientific theory. What a falsificationist-minded philosopher such as Popper might be inclined to see as a refuting instance of a theory, Kuhn interpreted as a puzzle that a modified form of a theory, derived from the same paradigm, could turn into a confirming instance.

Paradigm Change and Scientific Revolutions

Kuhn's paradigm concept provided him with a radically different approach to the problem of scientific change. Virtually all philosophy of science before Kuhn's book appeared subscribed to the notion of cumulative scientific change, in which there were occasional major breakthroughs, such as the transition from Newtonian mechanics to Einsteinian relativistic mechanics. Even such major breakthroughs, however, were conceived of as representing the continuous application of a common set of methods and standards. Kuhn's analysis, building on the more inclusive concept of 'paradigm,' proposed an account of scientific revolutions which highlighted discontinuity, noncumulativity, and even incommensurability between successive paradigms. In an all too brief characterization to provide the rich detail found in *The Structure of Scientific Revolutions*, let us step through the stages of paradigm development from "normal science" to scientific revolution.

A paradigm is an open-ended construct. It solves some major scientific problems but leaves for additional workers many tasks of extension, application, and discovery. These tasks are not seen as threatening the general claims of the paradigm—they constitute "puzzle solving" and represent, in Kuhn's term, "normal science." Many years and much scientific work can and has taken place under this description. The scientific contributions are important ones, and the best practitioners of normal science are accorded various honors such as election to distinguished scientific academies and the receiving of Nobel prizes.

As scientific investigation continues, some puzzles become seen as more resistant. Other unanticipated discoveries may be made that do not fit well with the elements of the dominant paradigm. These recalcitrant, paradoxical phenomena become *anomalies* from the perspective of the paradigm. As such anomalies increase or are perceived to deepen, that field of science can enter a *crisis*—a state in which *extraordinary science* becomes more legitimate, in which fundamental assumptions of the paradigm can be challenged, and alternative paradigms are made more welcome in the scientific literature and at scientific meetings. Ultimately, a new paradigm which solves some of the anomalies will appear, and attract a number of scientists to practice science according to the assumptions, methods, and standards of the new paradigm. Kuhn suggested these proponents of the new paradigm will typically be young scientists whose training in and stake in the older paradigm are minimal.

This is a "scientific revolution" but one in which the old paradigm is not seen as a limiting case of the new paradigm; rather, scientists accepting the new paradigm have in a sense shifted to a different type of world with a different language from that of the earlier paradigm. Kuhn frequently cites various examples of gestalt figure-ground perceptual shifts as metaphors for what happens in paradigm shifts. I mentioned earlier that I believe it was Kuhn's new conception of a *paradigm* and his association of a *shift of paradigms* with dramatic discontinuity and irrationality in science that has most captured the imagination of nonphilosophers. The relation of successive (or competing) paradigms was characterized by Kuhn (and by Feyerabend) as one of *incommensurability*. The problem of incommensurability is of interest to us in connection with the evaluation and acceptance of scientific theories, since it implies that intertheoretical standards of evaluation and comparison are impossible. The impossibility of such standards has three sources: the theory-ladenness of criteria, of scientific terms, and of observation terms.

The Theory-Ladenness of Criteria

I have already noted earlier that in his book, *The Structure of Scientific Revolutions*, Kuhn suggested that the standards by which groups of scientists judge theories are heavily conditioned by the theory itself, or, in Kuhn's term, by the *paradigm*. This is because a paradigm is *broader* than a theory and contains within it not only the abstract theory per se but also metaphysical commitments, methodology and rules, specific examples of application, *and* standards of evaluation.

In his 1970 Postscript Kuhn added that:

> Judgments of simplicity, consistency, plausibility, and so on often vary greatly from individual to individual. What was for Einstein an insupportable inconsistency in the old quantum theory, one that rendered the pursuit of normal science impossible, was for Bohr and others a difficulty that could be expected to work itself out by normal means. . . . In short, . . . the application of values is sometimes considerably affected by the features of individual personality and biography that differentiate the members of the group.[1]

Such variation is important in permitting creativity to function, leading to the pursuit of new theories. It is interesting to note, however, that though such variation does occur, I believe that there is not as much as Kuhn suggests, and as time passes this variation converges toward a mean among scientists, but this is better related to points to be taken up later on in this chapter.

The Theory-Ladenness of Scientific Terms

Both Kuhn (and Feyerabend) have argued that prima facie cases of intertheoretic comparability are illusory because of the radically different meanings that scientific terms in different theories possess. Thus, though it appears that Newtonian mechanics is a limiting case of Einsteinian mechanics (for velocities much less than light), Kuhn asserts that what one obtains in such a limiting case is *not* Newtonian theory but at best an explanation as to why Newton's theory has worked at all. A term such as "mass" in Einsteinian theory has associated with it the ideas of variability with relative velocity and intercovertibility with energy (expressed by the famous equation $E = Mc^2$). These notions are foreign and inconsistent with the concept of Newtonian "mass," and, accordingly, any identity, even under restricted conditions such as $v \ll c$, is incorrect and misleading. In the area of the biomedical sciences a similar problem arises with different concepts of the classical and molecular characterization of the gene, a position which has been forcefully argued by Hull,[12] Rosenberg,[13] and others.

The Theory-Ladenness of Observations

Kuhn (and Feyerabend also) asserts that observations or experimental results, which a scientist might be disposed to think would be independent of theory, are actually heavily conditioned by theory. The position of Kuhn (and Feyerabend) in this regard can be set forth by noting that they do not subscribe to the thesis that there is a theory-neutral observation language. Thus, basic meterstick and volume measures, pointer readings, solution colors, sedimentation coefficients, as well as, I would surmise, the measurement of peripheral nerve conduction velocities and various "endpoints" and "surrogate endpoints" in clinical research, are viewed as meaningless per se, and as noncommon to proponents of two inconsistent theories. Any evidential weight of observational results, on this view, is dependent on the assumption of the truth of the theory explaining those results.

There are several instances in the current biomedical science which offer prima facie support for this view of the theory-ladenness of observational reports. For example, data regarding the sequence of amino acids in immunoglobulins were interpreted significantly differently by proponents of the germ-line and somatic mutation theories to explain the source of antibody diversity (see Burnet[14] and Hood and Talmage[15]). One of the issues that should be considered in this book is the extent to which experimental results are significantly conditioned by assumptions related to different paradigmatic approaches.

These three aspects of the incommensurability problem have led Kuhn (and Feyerabend) to assert that it is impossible to compare and evaluate paradigms and very general theories disinterestedly and objectively. Proponents of different general theories live in "different worlds," worlds which are noncomparable or "incommensurable" with one another. Such a position, though it does explain scientific inertia and the resistance of scientists to give up their preferred theories, raises the specter of subjectivism in science (see Scheffler[16] for an early critique along these lines).

Kuhn's Disciplinary Matrix

In a reaction to some strong criticism by philosophers of science, in his 1970 Postscript Kuhn reconsidered his notion of a paradigm. Shapere's critique of the paradigm concept was both typical and influential. Shapere wrote that Kuhnian scientific relativism,

> while it may seem to be suggested by a half-century of deeper study of discarded theories, is a *logical* outgrowth of conceptual confusions, in Kuhn's

case owing primarily to the use of a blanket term. For his view is made to appear convincing only by inflating the definition of "paradigm" until that term becomes so vague and ambiguous that it cannot easily be withheld, so general that it cannot easily be applied, so mysterious that it cannot help explain, and so misleading that it is a positive hindrance to the understanding of some central aspects of science; and then, finally, these excesses must be counterbalanced by qualifications that simply contradict them.[17]

Masterman[18] similarly discerned some 21 different senses of the term "paradigm." Kuhn himself admitted in the second edition of his *Structure* . . . that there were "key difficulties" with the concept of a paradigm, and proposed replacing it with the term "disciplinary matrix." The paradigm notion had, he proposed, two interpenetrating but still rather different aspects, and Kuhn proposed that he use some new terminology, specifically the idea of a 'disciplinary matrix,' to clarify the earlier term 'paradigm.' The first aspect of 'paradigm,' which we may describe as a disciplinary matrix per se, was similar to what traditional philosophers of science had understood a scientific theory to be; namely, collections of (1) symbolic generalizations and (2) models. In his 1962 book Kuhn had made a significant departure from traditional philosophy of science and had introduced "values" — primarily of an epistemic sort — as an important component of a paradigm. This intrinsic value component is again reintroduced as Element 3 of the disciplinary matrix in his 1970 edition. A fourth component of the "disciplinary matrix," which Kuhn termed an "exemplar," was, however, viewed by Kuhn as a *distinctly different sense of paradigm*. Kuhn wrote:

> For . . . [the fourth sort of element in the disciplinary matrix] the term 'paradigm' would be entirely appropriate, both philologically and autobiographically; this is the component of a group's shared commitments which first led me to the choice of that word. Because the term has assumed a life of its own, however, I shall here substitute 'exemplars.' By it I mean, initially, the concrete problem solutions that students encounter as part of their scientific education, whether in laboratories, on examinations, or at the ends of chapters in science texts. To these shared examples should, however, be added at least some of the technical problem-solutions found in the periodical literature that scientists encounter during their post-educational careers and that also show them by example how their job is to be done. More than other sorts of components of the disciplinary matrix, differences between sets of exemplars provide the community fine-structure of science.[1] (186–187)[my emphasis]

KUHN'S MOST RECENT VIEWS ABOUT SCIENTIFIC PROGRESS

The specter of subjectivism has troubled Kuhn, and even in his 1962 writing he wrestled with the implications of his views for scientific progress. The problem arises from the all-inclusive character of a paradigm and the inability — from a Kuhnian perspective — to find any external "Archimedean" point, as it were, from which to judge the adequacy of competing paradigms. Since there are no objective extra-paradigmatic standards of truth, a number of scholars have seen special interests including power, politics, and authority as constituting the major forces for paradigm success or failure. Kuhn, in his most recent writing is critical of this interpretation of his account. In his 1992 lecture he cites the movement in science studies known as "the strong program," which he describes as holding that:

. . . power and interest are all there are. Nature itself, whatever that may be, has seemed to have no part in the development of beliefs about it. Talk of evidence, of the rationality of claims drawn from it, and of the truth or probability of those claims has been seen as simply the rhetoric behind which the victorious party cloaks its power. What passes for scientific knowledge becomes, then, simply the beliefs of the winners (pp. 8-9).[2]

Kuhn hastens to add that he is "among those who have found the claims of the strong program absurd: an example of deconstruction gone mad." (p. 9)

Kuhn continues to struggle with how one might account for scientific progress in the absence of any extra-paradigmatic standards of validity. Reprising a theme from his 1962 book in his 1992 lecture, he argues for a kind of Darwinian concept of scientific progress in which we can see *from* where we have come, but not *toward* what point we are going. Though disagreeing with any possibility of accessing an objective scientific reality, Kuhn believes that some type of sense can be made of comparative evaluation and of a notion of the rationality of incremental change, and suggests that focusing on these themes will provide a place for evidence and reason in analyses of scientific change. It remains to be seen, however, — perhaps in his forthcoming new book that will address these issues — exactly how even comparative evaluation and incremental changes can be provided with a common base that is not idiosyncratic to specific paradigms.

The other major theme to be found in Kuhn's 1992 lecture calls for

abandoning the view of science as a single monolithic enterprise, bound by a unique method. Rather it should be seen as a complex but unsystematic structure of distinct specialties or species, each responsible for a different domain of phenomena, and each dedicated to changing beliefs about their domain in ways that increase the accuracy and other standard criteria [of comparative evaluation] I've mentioned. For that enterprise, I suggest, the sciences, which must be viewed as plural, can be seen to retain a very considerable authority.(pp. 18-19)

The value of specialization, however, may have its drawbacks with respect to rationalizing scientific change, for Kuhn also notes that increased specialization can lead to "incommensurability" between the proponents of two different specialties. (p. 20)

SPECIALIZATION AND METHODOLOGICAL PROBLEMS IN ENVIRONMENTAL SCIENCE

I want to close this chapter by following up on some of Kuhn's suggestions regarding *differences* between scientific specialties, and in particular examine whether there are some differences between the physical sciences — from which virtually all of Kuhn's examples are drawn — and the biological and clinical sciences that serve as the basis of environmental science.

Theory and Experiment in Biology Compared and Contrasted with Physics

Biology in general, and environmental biology as well, share a number of common methodological assumptions with all the natural sciences, including the physical sciences. The life sciences, however, also possess some special methodological features which it will be useful to distinguish. Because each of the natural sciences seeks reliable general knowledge, the canonical scientific methods, such as the methods of agreement and difference, are widely employed in the life and nonlife sciences. These methods can be thought of as attempting to discern the causal structure of the world, and in their application scientists endeavor to identify possible confounding factors which can lead to spurious inferences about causes and effects. Thus all natural scientists attempt to control for interfering and extraneous factors, frequently by setting up a control comparison or a control group. Such controls are a direct implementation of what John Stuart Mill termed the method of difference and Claude Bernard the method of comparative experimentation, discussion of which can be found elsewhere.[19]

In the biological sciences, including environmental biology, some added complexity is frequently encountered due to biological diversity and the number of systems which strongly interact in living organisms. It is the backdrop of evolutionary theory that allows us to understand why there can be both extensive and subtle variation in organisms and mechanisms, as well as why there may be narrowly defined precise mechanisms that are (nearly) biologically universal, such as the genetic code. Variation due to meiosis, mutation, and genetic drift, for example, predict that extensive variation should occur most frequently in evolving populations where strong selection pressures toward precise and universal mechanisms are not present. Alternatively, where variations would almost certainly be lethal, there exist strong pressures toward the fixation of (nearly) universal mechanisms. Thus evolutionary theory at a very general level explains some of the specific and general features of other theories and models in microbiology.[20]

Because broad and subtle variations may be encountered in biology, special attention frequently needs to be given to ensuring the (near) identity of the organisms under investigation, except for those differences which are the focus of the scientist's inquiry. Accordingly, the development of special strains of organisms and the identification and classification of populations (and subpopulations) assumes an urgency that frequently can be ignored in the physical sciences, where, for example, all electrons are identical. It is in connection with the satisfaction of these urgent needs that the development of specialized techniques and statistical methodologies that are sensitive to such complexity and interactions becomes so vitally important.

Organism and subsystem variation does not only influence the experimental arena, but also figures in what constitutes the biological analog of "theory" as well. Genetic and environmentally produced variation frequently results in the need of biologists to focus on "model" organisms and on prototypical systems, which are highlighted against a backdrop of similar but different organisms and mechanisms. I think it is here that Kuhn's notion of the *exemplar* dimension of a disciplinary matrix has an important role to play, because it reminds us that exemplars — think of these as model systems in biology — are the focus of much research and of differences between scientific communities who utilize different exemplars and often obtain different biological and clinical results.[21] The amount of variation and the numbers of levels of approach in biology are extraordinarily broad.

Because of the diversity of model organisms and approaches, biologists often find themselves practicing what a 1985 National Academy of Sciences report called "many-many modeling" in a complex "biomatrix."[22] The report, which represented the results of a series of workshops directed by Harold Morowitz, introduced a notion of the "matrix" of biomedical knowledge: "The workshops demonstrated that the results of biomedical research can be viewed as contributions to a complex body, or matrix, of interrelated biological knowledge built from studies of many kinds of organisms, biological preparations, and biological processes at various levels." From within this multidimensional matrix, "many-many modeling" occurs, in which analogous features at various levels of aggregation are related to each other across various taxa. The report notes: "An investigator considers some problem of interest — a disease process, some normal physiological function, or any other aspect of biology or medicine. The problem is analyzed into its component parts, and for each part and at each level, the matrix of biological knowledge is searched for analogous phenomena . . . Although it is possible to view the processes involved in interpreting data in the language of one-to-one modeling, the investigator is actually modeling back and forth onto the matrix of biological knowledge."

CONCLUSION

It strikes me that these issues of methodology, controls, complexity, and many-many modelings represent at least part of the focus that the proponents of various approaches to environmental risk analysis need to address. Kuhn's classical analysis of the structure of science as consisting of paradigms, and the difficulty of rationally explicating paradigm shifts, has led Kuhn himself to suggest that we should focus on the methods of the special sciences, and attempt to understand scientific change as incremental, but largely rational shifts. This will not be an easy task because of the complexity of the subject matter the environmental sciences needs to address, and because we also realize — post-Kuhn — that power and special interests also have major roles to play in scientific change. Arriving at rational judgments in this field will thus be doubly difficult, but it is an enterprise that is well worth the effort.

REFERENCES

1. Kuhn, T.S. *The Structure of Scientific Revolutions*, 1962; 2nd ed. (Chicago, IL: University of Chicago Press, 1970).
2. Kuhn, T.S. "The Trouble with the Historical Philosophy of Science." The Robert and Maurine Rothschild Distinguished Lecture, 19 November, 1991. Department of the History of Science, Harvard University, Cambridge, MA.
3. Giere, R. *Explaining Science: A Cognitive Approach* (Chicago, IL: University of Chicago Press, 1988).
4. *The Washington Post*, February 28, 1993, p. C1.
5. Feyerabend, P.K. "Explanation, Reduction, and Empiricism," in *Minnesota Studies in the Philosophy of Science*, Vol. 3, H. Feigl and G. Maxwell, Eds. (Minneapolis, MN: University of Minnesota Press, 1962), p. 28.
6. Feyerabend, P.K. "Consolations for the Specialist," in *Criticism and the Growth of Knowledge* (Cambridge, England: Cambridge University Press, 1970), p. 197.
7. Kuhn, T.S. *The Copernican Revolution* (Cambridge, MA: Harvard University Press, 1957).
8. Hanson, N.R. *Patterns of Discovery* (Cambridge, England: Cambridge University Press, 1958).
9. Toulmin, S. *Foresight and Understanding* (London: Hutchinson, 1961).
10. Hull, D. *Darwin and His Critics* (Cambridge, MA: Harvard University Press, 1973).
11. Schaffner, K.F. "Theory Change in Immunology: The Clonal Selection Theory — Part I: Theory Change and Scientific Progress; Part II: The Clonal Selection Theory," *Theoretical Medicine*, 13(2):175–216 (1992).
12. Hull, D. *Philosophy of Biological Science*, (Englewood Cliffs, NJ: Prentice-Hill, Inc., 1974).
13. Rosenberg, A. *The Structure of Biological Science* (Cambridge, England: Cambridge University Press, 1985).
14. Burnet, F.M. "A Darwinian Approach to Immunity," *Nature* 203:451 (1964).
15. Hood, L., and D. Talmage, "Mechanisms of Antibody Diversity: Germ Line Basis for Variability," *Science* 168:325 (1970).
16. Scheffler, I. *Science and Subjectivity* (Indianapolis, IN: The Bobbs-Merrill Company, Inc., 1967).
17. Shapere, D. "The Structure of Scientific Revolutions," *Philosophical Review*, 73:393 (1964).

18. Masterman, M. "The Nature of a Paradigm," in *Criticism and the Growth of Knowledge*, I. Lakatos and A. Musgrave, Eds. (Cambridge, England: Cambridge University Press, 1970), p. 59.
19. Schaffner, K.F., "Philosophy of Method," in *Encyclopedia of Microbiology*, Vol. 3, J. Lederberg, Ed. (San Diego, CA: Academic Press, Inc.), p. 111.
20. Schaffner, K.F., "Theory Structure in the Biomedical Sciences," *Journal of Medicine and Philosophy*, 5:57 (1980).
21. Schaffner, K.F. "Exemplar Reasoning About Biological Models and Diseases: A Relation Between the Philosophy of Medicine and Philosophy of Science," *Journal of Medicine and Philosophy*, 11:63 (1986).
22. Morowitz, H. *Models for Biomedical Research: A New Perspective* (Washington, DC: National Academy of Sciences, 1985).

CHAPTER 2

A Brief History and Critique of the Low Dose Effects Paradigm

Leonard Sagan, Electric Power Research Institute,
Palo Alto, California

> Man has such a predilection for systems and abstract deductions that he is ready to distort the truth intentionally, he is ready to deny the evidence of his senses in order to justify his logic.[1]
>
> Risk assessment data can be like the tortured spy. If you torture it long enough, it will tell you anything you want to know.[2]
>
> Knowledge is like a ship because once it is in the bottle of truth it looks as though it always has been there and it looks as though it could never get out again.[3]

Many members of the public are frightened of even trivial exposures to environmental chemicals and radioactivity and are therefore willing to support the expenditures of large sums of money to reduce those exposures. While some scientists decry this exaggerated fear, the public is justified in their fears, not because the fears are based on demonstrated risks, but because many of those same scientists have themselves promulgated theories upon which those fears are based. The theory which I have in mind is that if harm is demonstrated at very high doses, then even very small exposures are treated as though they are harmful; i.e., the no-threshold model.

Those public fears of small environmental exposures have created a paralysis of environmental policy. For example, no waste repository for the medical uses of radioactivity can be sited, threatening to shut down the use of medical radioisotopes. The Department of Energy (DOE) has embarked on a program of radiation cleanup at DOE facilities which is anticipated to cost as much as 200 billion dollars. Yet, studies show that radiation exposures to the public will be reduced only by trivial amounts and human health will benefit not at all. Our chickens have come home to roost.

Another example: tens of billions of dollars have been spent in the "cleanup" of chemical waste sites without any persuasive evidence that human health has benefited.

How did this no-threshold model develop? It originated from the difficulty or impossibility of detecting the very small effects, whether harmful, beneficial, or null, which may result from low levels of exposure. Out of the need to regulate, and out of a sense of what at the time appeared to be prudence, the assumption was made that very low exposures are harmful at any level, no matter how small. This model, or paradigm, became widely adopted in the 1970s by regulators, but also came to be accepted as established truth by the public and by scientists themselves.

How is it that scientists would buy into a model for which there was little evidence? The popular view of scientists is that they are cold, aloof, dispassionate, and free of social or political values. Similarly, the popular view of scientific knowledge is that it is also objective and value free. The thesis asserted here is that, particularly when there is great scientific uncertainty, as is true of risks from low environmental exposures, social and political ideology will influence the interpretation of science.

ACCEPTED MODELS, OR "PARADIGMS"

In his book, *The Structure of Scientific Revolutions*, Thomas Kuhn alleged that most scientific thinking is dominated by certain sets of assumptions or models, developed as explanations of observed phenomena.[4] He called these "paradigms," and observed that scientists working in the field adopt the paradigm unthinkingly, never challenging the underlying assumptions, and are in fact more likely to attack challengers than to question the paradigm itself; i.e., scientists are essentially a conservative lot.

Scientific models are constructs devised by scientists to explain observable phenomena. They are to be distinguished from facts in that facts are observable, and under specified conditions can easily be replicated, somewhat as a cooking recipe can be replicated, by other interested scientists. Many of the assumptions built into models, however, cannot be replicated; they may be reasonable guesses about how things work, but they cannot be observed; they are not "facts." For example, it was once assumed by scientists on the basis of their observations that the Earth is flat and that the Earth rotates around the sun. That was a reasonable model for many hundreds of years. It fit the available facts. Ultimately, that model was abandoned when facts became available which were inconsistent with the model. Another model was adopted, one in which the Earth was round and rotated around the sun. Now, the interesting thing about all of this is that scientists themselves frequently ignore the difference between the facts and the assumptions built into a model or paradigm—the model which was developed as a useful tool becomes a universal truth.

Kuhn also alleged that information inconsistent with the accepted model is ignored and censored as heretical until contrary evidence becomes so strong that a "paradigm shift," or new model is adopted. New models emerge, not through gradual evolution, but through revolutionary change driven from outside of the "establishment," not from within; members of the establishment have too much invested, intellectually and economically, in the traditional model. Think of how the Church resisted the new model of the Earth rotating around the sun, and how Galileo was nearly excommunicated.

Kuhn assumed that the influence and operation of paradigms was peculiar to science. Barker has more recently pointed out that all areas of human activity, certainly including the business world, are controlled by paradigms.[5] Not everyone agrees with Kuhn;[6] his ideas have stirred a lively debate about the nature of science, and the nature of scientists.

HISTORICAL DEVELOPMENT—HOW DID WE GET HERE?

The observation that exposure to high amounts of ionizing radiation could produce harmful, even lethal, effects was recognized shortly after the discovery of the existence of ionizing radiation in 1895.

It was thought, however, that radiation effects obeyed a threshold response; that is, only high exposures which exceeded a threshold would produce biological effects. Occupational exposure standards were based upon such a presumption (the "old paradigm").

Historically, the common practice in setting occupational exposure standards for chemicals was to identify the lowest dose or concentration at which observed health effects occur. For the sake of prudence, the standard was then set at some appropriately lower level.[7]

Following the second World War, however, this strategy was reconsidered; a "paradigm shift" occurred. I believe that there were three reasons for this, one of which came from engineering, another from biology, and the third from social psychology.

While engineers, in designing for safety, had previously followed a strategy in which they calculated maximum loads or stresses and then added a safety factor, similar to the practice of toxicologists in setting chemical exposure standards, nuclear engineers in calculating the risks of nuclear releases rejected the notion of a threshold and of perfect safety. Accepting that there was no absolute safety, they estimated risks of accidental releases which were always finite, never zero. They then designed backup safety systems to contain or minimize the consequences of even remotely possible accidents. The successful containment of the accident at Three Mile Island illustrates the prudence of this practice. Nevertheless, the message to the public was that accidents would occur, and furthermore, that technology was inevitably associated with increased risk.

There was also biological evidence that challenged the older notion of a threshold. This evidence was based partly on the studies of mutagenesis conducted in fruit flies by Herman Muller (1890-1960), studies in which he was unable to demonstrate a threshold. Muller's concerns regarding wide-spread industrial uses of ionizing radiation led him to suggest that thresholds for genetic effects might not exist—no definitive thresholds had been demonstrated, or rejected.[8] Muller never ceased to warn physicians of the genetic effects of the use of radiation. Following the war, Muller delivered a lecture on the genetic hazards of nuclear testing to the National Academy of Sciences that attracted great attention. While Muller himself supported the development of nuclear weapons, his concerns about genetic effects became a focal point in the demands for a test ban treaty.[9] At about mid-century, radiation biologists were also developing a theory of radiation effects which presumed that those effects were the result of minute "hits," or damage to cells much like the hits of a bullet in a target. Indeed, the theory was known as "target theory." These hits would occur randomly, and so even the smallest dose would have some statistical probability of hitting the target and producing harmful effects. The effects which were then of greatest concern were threats to the gene pool. This assumption was based on studies, such as those of Muller, which showed the mutagenic potential of radiation. On the basis of prudence, then, public policy authorities in the 1950s adopted a policy in which it was assumed that even very low exposures of radiation might be harmful.

At about the same time (the 1950s) it was also recognized that mutagenesis was often an important step in the process of carcinogenesis. This recognition was used to support the practice of assuming that even low exposures could be carcinogenic. Subsequently, as genetic research on mammalian species demonstrated that the risk of mutagenesis found in fruit flies had probably exaggerated risks to humans, the threat of cancer became the predominant concern of the radiation protection community, particularly as cancers other than leukemia began to appear in significant numbers in the survivors of the atomic bombings.

Something else happened in the 1960s that contributed still further to the concern about contamination of the environment, and that was improvement in the lower detection limits of chemical analysis.

THE CULTURE OF THE '60s—THE NEW ENVIRONMENTALISM

The 1950s and '60s were a period during which the public was becoming increasingly aware of environmental pollution with industrial chemicals. Rachel Carson's book, *Silent Spring*, was a milestone in arousing public concern.[10] Carson emphasized not only ecological consequences of environmental pollution but also specifically indicted environmental chemicals as important human carcinogens. She had brought to public attention observations on the effects of pesticide residues on the fertility of birds; it required only a small leap of faith to believe that environmental contaminants could also produce human health effects.

Something else was going on in the '50s and '60s that had a powerful effect on the public view of environmental radioactivity, and that was the great debate on weapons fallout. Those who were very much opposed to weapons testing emphasized the dangers of fallout to human health, even at very low levels. These estimates of disease were not based on observations of disease, but rather on extrapolations from high level exposures. Professor Ernest Sternglass of the University of Pittsburgh criss-crossed the country, reporting on his studies showing that thousands of babies were being killed by fallout. Those studies were patently flawed, obvious attempts to exploit a scientific gloss for political purposes, yet there was precious little response from the scientific community, which seemed satisfied to sit on the sidelines.

Whatever the reasons, it is remarkable to look back at the scientific literature of the 1960s and '70s and find almost no resistance to the no-threshold model. The reason is that it suited everyone's purposes. The radiation protection community benefited greatly from the increased fears of low doses of radiation. Radiation researchers benefited from increased fear and the consequent increased funding of radiation research. Lawyers benefited from increased litigation resulting from the public conviction that low levels of exposure had caused cancers. Regulators certainly had their lives made easier and their budgets enhanced by the adoption of the linear no-threshold model. But aside from special interest groups, did society really benefit when it was led to believe that something is true that remains unknown? The economic and social costs, and the political and environmental problems we have engendered by tacit acceptance of the no-threshold paradigm have not been quantified, but are undoubtedly enormous.

ARE SCIENTISTS INFLUENCED BY VALUES?

If there is uncertainty regarding low dose effects, why did we choose the paradigm that we did? I will suggest that when science is uncertain, values take over. My assertion is that the environmental movement benefited from the convenient but unproved assumption that environmental contamination posed an important health threat, and that scientists who shared those environmental concerns were perfectly happy to participate by providing risk estimates at levels of exposure below those where harm could be demonstrated.

Is it really possible that our no-threshold paradigm was an invention developed to satisfy the moral demands of society? To answer that question, let me first describe what most people think of the scientific method. Scientists are seen as those who operate in a value-free world, searching in a neutral way for an objective truth. They observe the world dispassionately, collecting data in a scrupulously objective fashion, which they then dutifully report in peer reviewed journals. Those reports then become the substance of an ever expanding knowledge.

In fact, scientists themselves select the cells, tissues, or animals which are most likely to produce the desired results. They are very likely to select and interpret their data in such a way as to support their theory. Papers are then subject to the judgment of an editor, who has his own judgment of what is desirable to publish.

While unconscious bias may easily enter this process at any point, conscious bias (cheating, lying) may also be introduced. As a consequence, medical and scientific history is replete with examples of scientists who either consciously or unconsciously interpreted their data to support theory, and there are an equal number of examples of theory developed to prove social and economic theory. This is not the place for an extensive exposition, but one example might do: As elegantly illustrated by Stephen Jay Gould, Dr. George Morton finagled his data to support the theory, popular at the time, that the intelligence of blacks was inferior to that of whites, and the intelligence of Indians was intermediate.[11] The remarkable thing is that he was scrupulous in collecting his data, which he published. His unconscious bias was so strong that he did not recognize his biases in the interpretation of the data, allowing Gould to expose those biases a hundred years later.

Now I would not like to suggest that a paradigm persists only because it is suitable to current value systems; there are other reasons as well. One is that, just as the illusion of power increased the stature of the medicine man and shaman, the illusion of precise knowledge lends prestige to the scientific community. For this reason, scientists are reluctant to admit how little they know.

RISK ASSESSMENT

Given a paradigm in which it is asserted that even very low exposures of radiation are damaging, there developed the practice, known as risk assessment, in which the consequences of very low exposures could be calculated, based on observations at high exposures. Such an assumption had the marvelous benefit that it became childishly easy to estimate risks at low exposures.

It also became easy to calculate aggregate risks in exposed populations and to use these for political purposes. For example, when Willard Libby calculated that the cancer risk from weapons fallout was one in a million, Linus Pauling, arguing on the other side of the weapons test ban issue, then concluded that in the world population of 4 billion, one in a million would be a total of 4,000 cases of cancer.[9] The political implications of the small risk of one in a million and the large number of cases that would result from that risk in a large population, 4,000, are obvious.

In 1957, E.B. Lewis published an article in *Science* in which he calculated the proportion of leukemia cases occurring in the United States which could be attributed to background ionizing radiation, assuming that no thresholds existed.[12] The article created a practice which persists to the present time and is widely viewed in the scientific community as having practical validity.

The use of this risk model is now widely applied and explicitly accepted as the truth. The model is applied with great precision. For example, the British calculated 12.7 cases of leukemia in the United Kingdom as a result of the Windscale accident in 1957. How was this number arrived at? First, by making crude estimates of the very low exposures to individuals in the population, then multiplying by the large number of persons exposed throughout England, then by making the assumption of the absence of a threshold and assuming that even at trivial doses to individuals, effects nevertheless occur. Given that these cancers would, if they did occur, be dispersed among the hundreds of thousands that would occur normally, it is clearly impossible to ascertain whether 12.7 cases occur or not.

The same assumptions are made by the Environmental Protection Agency in estimating that 15,000 lung cancer deaths occur in the United States each year as a result of exposure to residential radon. Similar risk estimates are conducted for trace exposures to chemicals in the food supply or air.

As noted by Ehrenfeld, "We believe implicitly in our models. The more specific their predictions are, the more we believe in them, no matter how scientifically preposterous and absurd that specificity is."[13]

IS THE PARADIGM OBSOLETE?

On scientific grounds, we have now moved considerably beyond the simplistic model of cancer as originating from a single exposure to an environmental mutagen, and beyond the assumption that industrial chemicals or radioactivity are important sources of mutagens in the diet.

Bruce Ames, Professor of Biological Chemistry at the University of California, Berkeley, once an outspoken critic of environmental pollution, has taken a leading role in challenging the notion that industrial agents in the environment are an important source of mutagens, not because they do not exist, but because their concentrations are low compared to those which occur naturally. He notes that all plant materials contain natural pesticides, as potent mutagenically as are industrial chemicals.[14] Many seasonings and spices are also known mutagens; examples are pepper and cinnamon. Furthermore, cooking, including baking and frying, add considerably to the burden of mutagens in the diet. The charring one sees on meats or toasted bread are bountiful sources of mutagens and proven carcinogens when applied in high concentrations in animal studies. So too is coffee (it's the roasting of the beans which is mostly responsible).

From these natural sources, the quantities of mutagens of natural origin in the usual diet dwarf the concentrations of mutagens represented by industrial pesticides by a factor of hundreds or thousands. Ames estimates that each day the average person consumes about 1500 milligrams of pesticides of natural origin compared with less than 0.1 milligram of synthetic pesticide. Unlike the older theories of mutagenesis which attribute ominous risks to each mutagenic event, we now know that damaging events to chromosomes occur very frequently, both because of exposure to environmental agents and because of the mutagenic effects of the body's own metabolic activities. Estimates are that each cell in the body is exposed to such possibly damaging events thousands of times per day. Fortunately, we now know, the body also has elegant mechanisms for repairing the great majority of the damage to the body's DNA. This repair mechanism declines with age, more rapidly in some than in others.

The efficacy of these repair mechanisms may be more important than exposure to mutagens in determining the growth of cancers. For example, a study conducted by Dr. Lawrence Grossman of the Johns Hopkins University shows that in a study of patients with basal cell cancer, a common variety of skin cancer, the ability of the repair mechanism to repair genetic damage is inversely correlated with the appearance of these cancers; i.e., decreased repair is associated with a higher risk of skin cancer.[15]

HORMESIS

Kuhn not only claimed that scientific thinking is dominated by paradigms, he also claimed that scientists are highly resistant to challenges to the transmitted paradigm, and vigorously resist challenges to the conventional wisdom. Information which is inconsistent with the paradigm, he said, is censored, not in the sense of an official or explicit censor, but in the sense that authorities such as journal editors and research-sponsoring agencies do not accept such research as legitimate.

In contrast with the no-threshold theory, considerable evidence exists of beneficial or stimulatory effects at low exposure levels. Qualitatively different effects at different exposure levels should not cause any eyebrows to rise. After all, our common experience is replete with such examples. While a bottle of gin taken at one time may be lethal, a martini each evening actually appears to lengthen life. Many of the common vitamins, necessary to the diet, are harmful at high exposures. Sunshine in small exposures prevents rickets, at high exposures is carcinogenic. One could extend this list endlessly. It is possible that radiation is different from martinis and chemical exposures, but common sense would suggest that they are similar rather than different.

I frequently see in epidemiological or animal studies evidence of a protective effect at low exposures. These are always ignored by the authors who appear blind to these observations. For just one example, in a study of breast cancer among women who during the course of treatment for tuberculosis were regularly fluoroscoped, there is a distinct decrease in risk among those in a low dose category.[16] The author does not discuss it.

In 1979, Dr. T. Don Luckey published a book called *Radiation Hormesis* in which he gathered together the literature demonstrating exceptions to the general thesis that radiation is harmful at low exposures.[17] Indeed, the literature is full of reports suggesting that animals exposed to low exposures of radiation benefit from those exposures. Such benefits include enhancement of the immune system, increased resistance to infection, and increased longevity. Sagan has suggested several mechanisms which might explain how such effects could occur.[18]

There is also evidence that the original Mullerian theory of a decrease in fitness as a result of low dose radiation may be in error. John Gillespie, in reviewing the work of the geneticist, Bruce Wallace, describes how surprised Wallace was when he found that low dose radiation actually increased fitness. The experiment has now been replicated by others. Interestingly, even Wallace himself was unwilling to accept his own findings, and spent decades trying to reconcile his own work with the accepted paradigm.[19]

I am not arguing here that there is strong evidence that "a little radiation is good for you." Nor am I prepared to offer a new paradigm to replace the old. I am arguing that the evidence regarding the risks of low exposures is quite uncertain and that scientists and funding agencies should undertake the research necessary to produce the new paradigm.

I do not exclude the possibility that a little radiation may be both "good" and "bad," for different people, or even for the same person. I am also arguing that we act as though we know the answer to these questions, when in fact there is great uncertainty about this, and we are not doing the research necessary to resolve the matter because we are paralyzed by our subservience to the paradigm.

Just as with radiation, there are numerous reports in the literature which suggest that very small exposures of chemicals, generally thought of as harmful, have a stimulatory effect at low exposures. This literature has been reviewed by Edward Calabrese, a toxicologist at the University of Massachusetts.[20]

As predicted by Thomas Kuhn, suggestions that the paradigm might be in error have been censored. Not only is there little scientific interest in pursuing hormesis, there would undoubtedly be little interest among funding agencies which are themselves captives of the paradigm.

HOW EASY WILL IT BE TO SHIFT THE RADIATION PARADIGM?

Not easy at all. In addition to the intellectual commitment to the paradigm that most of us share, there are now many constituencies that thrive on that paradigm. There is the environmental community, the medical community, the regulatory community, and the legal community, to name just a few. Each of these derives enormous benefits from continued dominance of the paradigm and would lose to the same extent from a paradigm shift.

As Alan Barker points out, "New paradigms put everyone practicing the old paradigm at great risk. And, the higher one's position, the greater the risk. The better you are at your paradigm, the more you have invested in it. To change your mind is to lose that investment."[21]

SUMMING UP—WHAT IS WRONG WITH THE PARADIGM?

There are several serious problems with the use of the existing radiation paradigm. One is the absence of supporting scientific knowledge of the existence of risks in whole animals or humans at low exposures.

A second is the existence of contrary information suggesting that low exposures of radiation may be associated with health benefits, not risks ("hormesis").

The third problem is that risk estimates at low exposures are accepted by the media and the public as scientifically valid, and project the view that "even the lowest dose is harmful," whereas in fact, we do not have evidence as to whether such exposures are harmful, harmless, beneficial, or all of these. As a consequence, the costs of maintaining the paradigm are becoming enormous.

Still another problem is that the conventional paradigm no longer helps us solve problems. We have continually conducted larger and larger studies, in both animals and humans, without shedding any new light on the nature of risks from low level radiation exposure.

While radiation epidemiologists have been unable to detect harm or benefit from low level exposures, statisticians and epidemiologists have their own biases (often limited to searching for excesses of cancer), and my intuition is that those biases preclude the demonstration of a hormetic effect, if it exists. Also, the evidence from animal studies suggests an increase in longevity, rather than a protective effect against cancer. A careful review of longevity among low dose exposed populations has not been carried out.

A strategy more likely to be useful in shifting the paradigm is likely to arise from knowledge of mechanisms operating at low exposures. Knowledge of these mechanisms could then permit more focused epidemiological studies. The rapid rise of interest in understanding mechanisms of toxicity at the molecular level, rather than continued dependency on studies of animals exposed at high levels is a promising start in that direction.[22]

WHAT CAN SCIENTISTS DO?

The difficulty in separating facts and values guarantees that scientists' values will continue to affect public policy. How can this role be reconciled with traditional democratic ideals? Lowrance has suggested the following guidelines for scientists:

Recognizing that they are making value judgments for the public, scientists can take several measures toward converting an "arrogation of wisdom" into a "stewardship of wisdom."

First, they can leaven their discussions by including critical, articulate laymen in their group.

Second, they can place on record their sources of bias and potential conflicts of interest, perhaps even stating their previous public positions on the issue.

Third, they can identify the components of their decisions being either scientific facts or matters of value judgment.

Fourth, they can disclose in detail the specific basis on which their assessments and appraisals are made.

Fifth, they can reveal the degree of certainty with which the various parts of the decision are known.

Sixth, they can express their findings in clear jargon-free terms, in supplementary nontechnical presentations, if not in the main report itself."[23]

REFERENCES

1. Dostoevsky, F.H. *Notes From the Underground*, New York: Dover Press, 1992.
2. Ruckelshaus, W.D., quoted in Cohen, *News and Numbers*, Ames, IA: Iowa State University Press, 1989, p. 105.
3. Collins, H.M., *Changing Order: Replication and Induction in Scientific Practice*, London: Sage Publications, Inc., 1985.
4. Kuhn, T., *The Structure of Scientific Revolutions*, 2nd ed., Chicago, IL: University of Chicago Press, 1970.
5. Barker, J. A., *Discovering the Future: The Business of Paradigms*, St. Paul, MN: ILI Press, 1988.
6. Schleiffer, I., *Science and Subjectivity*, Indianapolis, IN: The Bobbs-Merrill Co., 1967; Laudan L., *Science and Relativism: Some Key Controversies in the Philosophy of Science*, Chicago, IL: The University of Chicago Press, 1990.
7. ICRP, "International Recommendations for X-Ray and Radium Protection," *British Journal of Radiology*, 7:695 (1934); Taylor, L. S., "The Origin and Significance of Radiation Dose Limits for the Population," presented at the AEC Scientific and Technical Symposium, August 17, 1973, United States Atomic Energy Commission, WASH-1336, Washington, DC. 20545; Kocher, D.C., "Perspective on the Historical Development of Radiation Standards," *Health Physics*, 61:519 (1991).
8. Carlson, E.A., *Genes: Radiation, and Society: The Life and Work of H.J. Muller*, Ithaca, NY: Cornell University Press, 1981.

9. Hewlett, R. G., and J.M. Holl, *Atoms for Peace and War*, Berkeley, CA: University of California Press, 1989.
10. Carson, R. L., *Silent Spring*, Boston, MA: Houghton Mifflin, 1962.
11. Gould, S. J., *The Mismeasure of Man*, Cambridge, MA: Harvard University Press, 1983.
12. Lewis, E., "Leukemia and Ionizing Radiation, *Science*, 43:965 (1957).
13. Ehrenfeld, D., "Environmental Protection: The Expert's Dilemma, Philosophy and Public Policy," School of Public Affairs, University of Maryland, 11:8 (1991).
14. Ames, B.N., and L.S. Gold, "Chemical Carcinogenesis: Too Many Rodents," *Proc. National Academy of Sciences*, 87:7772 (1990).
15. Grossman, L., *Proc. National Academy of Sciences*, Feb, 1993.
16. Miller, A.B., G.R. Howe, G.J. Sherman, et al., "Mortality from Breast Cancer After Irradiation During Fluoroscopic Examinations in Patients Being Treated for Tuberculosis," *New England Journal of Medicine*, 321:1285 (1989).
17. Luckey, T., *Hormesis with Ionizing Radiation*, Boca Raton, FL: CRC Press, Inc., 1980.
18. Sagan, L., "On Radiation, Paradigms, and Hormesis," *Science*, 245:574,621 (1989).
19. Gillespie, J., "The Burden of Genetic Load" (book review), *Science*, 254:1049 (1991).
20. Calabrese, E.J., M.E. McCarthy, and E. Kenyon, "The Occurrence of Chemically Induced Hormesis," *Health Physics*, 52:531 (1987).
21. Barker, J.A., *Discovering the Future: The Business of Paradigms*, St. Paul, MN: ILI Press, 1988.
22. Cohen S.M., and L.B. Ellwein, "Risk Assessment Based on High-Dose Animal Exposure Experiments," *Chem. Res. Toxicol.*, 5:742 (1992).
23. Lowrance, W.W., *Of Acceptable Risk*, Los Altos, CA: Wm. Kaufman Inc., 1976.

CHAPTER 3

Primer on BELLE

Edward J. Calabrese, School of Public Health,
University of Massachusetts, Amherst

INTRODUCTION

I became interested in the biological effects of low-level exposures as an undergraduate student enrolled in a laboratory-oriented course in plant physiology. An experiment designed to assess the effects of a synthetic plant growth retardant, phosfon, brought forth the unexpected observations that the peppermint plants were apparently stimulated rather than retarded in their overall vegetative growth. The professor indicated that this observation was unexpected but may have been the result of a mislabeling error or perhaps of some as yet unrecognized phenomenon. He asked if anyone were interested in following up on the observation. Thus, I then began a several-year investigation that centered around the effects of phosfon on peppermint. In a series of investigations assessing the effects of phosfon on the growth of peppermint in soil, high doses were indeed inhibitory whereas low levels were stimulatory, displaying what has become called the beta-curve. I had originally thought that the phosfon might have been transformed into a stimulatory agent by soil microbes so the experiments were repeated in a soil-less medium, using hydroponics. However, the same dose-response pattern displaying stimulatory growth at low doses and inhibitory growth at high doses was observed.[1]

The range of doses employed in the phosfon studies was extremely broad, covering some six orders of magnitude of dose in the soil experiments. Stimulatory responses were observed over a 1,000-fold range in the low dose zone. The estimated number of molecules administered to the soil in the single dose treatment ranged from 2×10^{12} to 2×10^{15} in the stimulatory range to greater than 2×10^{17} in the inhibitory range (Figures 3.1 and 3.2).

A literature search of the effects of phosfon revealed that other researchers had reported similar stimulatory/inhibitory dose-response relationships with phosfon and several commercial plant species such as zinnias, sunflower, and morning glory.[2-5]

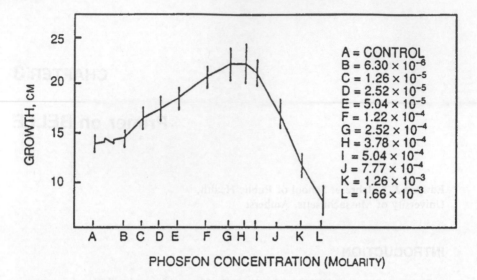

Figure 3.1.   The effect of phosfon (log scale) on the growth of *Mentha piperita* in a soil medium. Plants grown at concentrations of 1.26×10^{-5} *M* to 7.77×10^{-4} *M* were significantly taller ($p < 0.01$) than the controls. Plants grown at concentrations of 1.26×10^{-3} *M* and 1.66×10^{-3} *M* were significantly smaller ($p < 0.01$) than the controls. Plants grown at 6.30×10^{-6} *M* did not differ significantly from the controls. Vertical lines represent the standard error of the mean.[1]

While my direct research on the effects of phosfon on plant growth gave way nearly 25 years ago to a career change from the study of plants to animal model responses, the concept that chemical agents could affect biological systems in remarkably different ways depending on dose became strongly ingrained. Given this background, I was predisposed to be quite interested in the 1986 conference on Radiation Hormesis held in Palo Alto, California. That conference, directed by Dr. Leonard Sagan, provided the impetus that ultimately led to the start of the current BELLE initiative.

In May 1990 a group of scientists representing several federal agencies, the International Society of Regulatory Toxicology and Pharmacology, the private sector, and academia met to develop a strategy to encourage the study of the biological effects of low level exposures (BELLE) to chemical agents and radioactivity. The meeting was convened because of the recognition that most human exposures to chemical and physical agents are at relatively low levels, yet most toxicological studies assessing potential human health effects involve exposures to quite high levels, often orders of magnitude greater than actual human exposures. Consequently, risks at low levels are estimated by various means, frequently utilizing assumptions about which there may be considerable uncertainty.

The BELLE Advisory Committee is committed to the enhanced understanding of low-dose responses of all types, whether of an expected nature

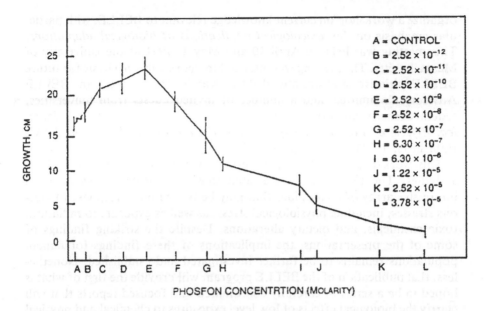

Figure 3.2.   The effect of phosfon (log scale) on the growth of *Mentha piperita* in a mineral nutrient solution. Plants grown at concentrations of 2.52×10^{-12} M to 2.52×10^{-8} M were significantly taller ($p < 0.01$) than the controls. Plants grown at 6.30×10^{-7} M to 3.78×10^{-5} M were significantly shorter ($p < 0.01$) than the controls. Plants grown at 2.52×10^{-7} M did not differ significantly from the controls. Vertical lines represent the standard error of the mean.[1]

(e.g., linear, sublinear) or of a so-called paradoxical nature. Paradoxical dose-response relationships might include U-shaped dose-response curves, hormesis, and, in some restrictive sense, biphasic dose-response curves. Although there are many scattered reports of such paradoxical responses in the biomedical literature, these responses have not generally been rigorously assessed, nor have the underlying mechanisms been adequately identified. Laboratory and regulatory scientists have tended to dismiss these paradoxical responses as curiosities or anomalies inconsistent with conventional scientific paradigms.

The focus of BELLE encompasses dose-response relationships to toxic agents, pharmaceuticals, and natural products over wide dosage ranges in in vitro systems and in vivo systems, including human populations. The initial goal of BELLE is the scientific evaluation of the existing literature and of ways to improve research and assessment methods. While a principal emphasis of BELLE is to promote the scientific understanding of low-level effects (especially seemingly paradoxical effects), the need to evaluate the risk assessment implications of such effects is also an important aspect of BELLE activities.

The BELLE Advisory Committee authorized Professor Edward J. Calabrese, School of Public Health, University of Massachusetts, Amherst, to organize a workshop on current knowledge relevant to BELLE, with particular emphasis on the *toxicological implications of biological adaptations*. This meeting was held on April 30 and May 1, 1991 at the University of Massachusetts. The meeting was intended to help establish a basis for future BELLE initiatives and was attended by seven invited speakers, the BELLE Advisory Committee, and a number of invited guests from universities, federal agencies, and private sector organizations. The proceedings (*Biological Effects of Low Level Exposures to Chemicals and Radiation*, Lewis Publishers, 1991) has been published.

This workshop provided an important benchmark for future BELLE activities. The presentations indicated that biological systems have an impressive array of adaptations that may be turned on in response to various stresses, including physiological stress, as well as exposure to radiation, toxic chemicals, and dietary alterations. Despite the striking findings of some of the presentations, the implications of these findings for human populations remains to be further investigated and established. Nonetheless, that publication of the BELLE program will provide the first of what is hoped to be a series of carefully coordinated and focused reports that will clarify the biological effects of low level exposures to chemical and physical agents on biological systems and human populations.

Hormesis: Its Historical Basis

The issue of how chemicals and radiation affect biological systems over wide dose ranges is not the purview of a single discipline but strongly cuts across many fields. In the experience of toxicology this area has been dominated over the past several decades by debates over the shape of radiation related dose-response relationships, and more recently via chemical responses as well.

The general premise of those who study the biological effects of chemicals and radiation is that all toxic responses observed at higher dose rates are the only effects elicited at lower exposure.[6] This central concept has served as the basis for how regulatory and public health agencies begin to deal with the process of deriving acceptable levels of exposures. A sizeable number of researchers have, however, suggested that at low dose rates paradoxical or U-shaped dose-response relationships have occurred, especially with respect to survival/longevity endpoints. In simple language, these observations are ones that don't seem to have followed the accepted dose-proportionality.

Whether these observations are the result of a "hitherto unsuspected truth,"[7] or the result of statistical artifacts is a critical issue. In fact, various investigators from multiple disciplines have been reporting such so-called anomalies since at least the 19th century, starting with the work of Hugo Schulz concerning the effects of chemicals on yeast fermentation.[8,9] He demonstrated that various toxic agents including mercuric chloride, iodine, bromine, arsenious acid, chromic acid, salicylic acid, and formic acid, could stimulate respiration and growth in yeast. This paradoxical mode of drug action ultimately became embodied in the so-called Arndt-Schulz law that "weak stimuli accelerate vital activity; medium ones promote it; strong ones inhibit it; and very strong ones snuff it out."

According to Boxenbaum,[6] there appeared to be no widespread adoption of the Arndt-Schulz Law since there was no serious reason for supposing that any such general law existed. Clark[10,11] argued that polyphasic response to drugs had been frequently seen, and there was no inherent explanatory power inherent in this observation.

The Arndt-Schulz Law nonetheless was supported by the research of Southam and Ehrlich[12] which noted that hot water soluble extracts of western red cedar heartwood (containing phenolic-type agents) displayed promotional effects on the growth of wood-decaying fungi incubated in malt-agar medium. At higher concentrations, the extract was either fungistatic or fungicidal, while at low concentrations enhanced fungal growth was seen in some species. The authors proposed that the enhanced growth was "an initial response, followed by progressive desensitization to sub-inhibitory concentrations of a toxic constituent of the extract." *These authors proposed the term hormesis to describe stimulatory (growth) affects induced by sub-inhibitory concentrations of toxic substances on organisms.* The etiology of the word is believed to come from a transliteration of the classical Greek, "hormetic" and "hormetikos" meaning "exciting, stimulating."[13]

The hormesis concept was enhanced further by Luckey[14] in 1956 when he demonstrated that germ-free birds (chicks and turkey poults) grew more rapidly when their diets were supplemented by low doses of antibiotics. Since their response could not be related to change in bacterial flora, Luckey proposed that the stimulatory action was caused by the low level exposure to the antibiotics. As a result, he offered the name "hormoligosis" (hormone = excite; oligo = small amounts) to describe compounds that were stimulatory at low doses, somewhat harmful at modest (therapeutic) levels, and toxic at high levels. Three years later, Luckey[15] had proposed a series of terms:

1. hormology—the effects of stimulants and excitation.
2. hormesis—stimulatory action of sub-inhibitory amounts of a toxin.
3. hormologosis—stimulation by small amounts of any agent upon living organisms.

Luckey became a strong advocate of the hormetic perspective in a series of papers demonstrating that the hormology concept was a generalized phenomenon producing a wide array of effects with an impressive variety of physical and chemical agents in plant and animal species.[16-18] Then, in 1975 Luckey broadened the term hormesis to include the action of any agent which is stimulatory at low doses.[19] According to Boxenbaum,[6] hormesis has become a catchall for any phenomenon deemed beneficial which can be induced by a low-level exposure of an otherwise toxic agent.

If Hormesis Exists, How Is It Recognized?

Stebbing[20] has presented several types of dose-response relationships that are purported to describe the hormetic response (Figure 3.3). The most frequently claimed response is the B-curve, and it is given greatest attention

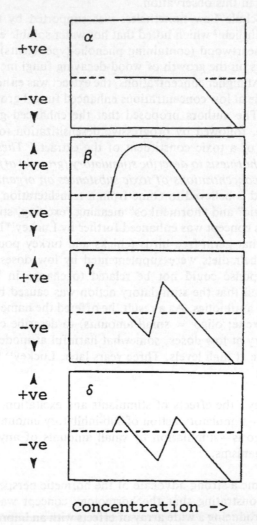

Concentration ->

Figure 3.3. Various types of concentration-response curves identified by Townsend and Luckey.[18] The α curve is the one generally found in growth experiments with toxicants, but other types of curve showing hormesis are also known. However, the most common form of hormetic curve is the β-curve.

here. In considering the B-curve, let us review a series of eight theoretical toxicological data sets (Figure 3.4): A and B have four data points, C—G have five data points, while H has 11 data points.

In order to convincingly fit a B-curve model, adequate data are necessary to both detect low dose stimulation and high dose inhibition. While data sets A—G are consistent with a hormetic hypothesis, they also leave considerable room for other models with which they may be comparably consistent. Unfortunately, most whole animal toxicological studies have limited experimental groups and have important limitations addressing the hormetic hypothesis. Figures 3.5 to 3.7 display widely cited data of Stebbing[20] which is used to support the hormetic hypothesis. These data points offer critical features that may define the characteristics of hormetic responses.

First, consider the range of responses: of the 32 examples presented by Stebbing[20], the maximum hormetic response (i.e., height of curve) ranged from 15% to 400% above the control, with the median increase being approximately 50% above the control. Five of the 32 values equaled or exceeded 2-fold above the control, while 11 are ≤ 1.25-fold greater than the control.

Second, the difference between the dose observed to cause the maximum stimulation and the dose estimate where the B-curve descended through the control value for 20 experiments using 12 pesticides on the growth of house crickets had a median of 3-fold, while a median of 4-fold was obtained from the data of Stebbing on 27 different experiments with a wide capacity of models, endpoints, and chemical agents.

Third, the ratio between the dose that caused 100% mortality in the insect and the peak growth stimulation response averaged 1/84 of the LD_{100} with a range of 1/10 to 1/1,000.

Fourth, the range of the stimulatory phase of the hormetic curve is difficult to discern, being the product of the quality and number of data points in the stimulatory range. Nonetheless, the data suggest that the hormetic response can range from several-fold to several orders of magnitude, with the majority approaching a 10-fold range.

Does the Hormetic Curve Always Take the Form of a B-Curve?

While the literature suggests that hormetic dose-response relationships often display a B-curve with respect to growth and longevity, hormetic-like mechanisms may also be operating to affect dose-response relationships that resemble hockey stick shaped dose-response relationships. For example, the dose-response relationship for CCl_4 induced liver damage at 48 hours after exposure in the Sprague-Dawley (SD) male rat displays a nonlinear dose-response relationship (Figure 3.8). However, if the early phase regeneration (EPR) hormetic response is ablated via prior treatment with colchicine, the dose-response relationship becomes more linear. Conse-

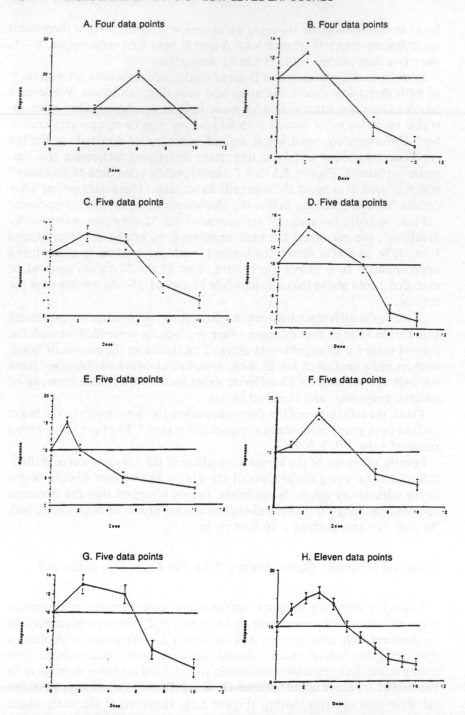

Figure 3.4. Hypothetical dose response relationships consistent with the hormetic hypothesis.

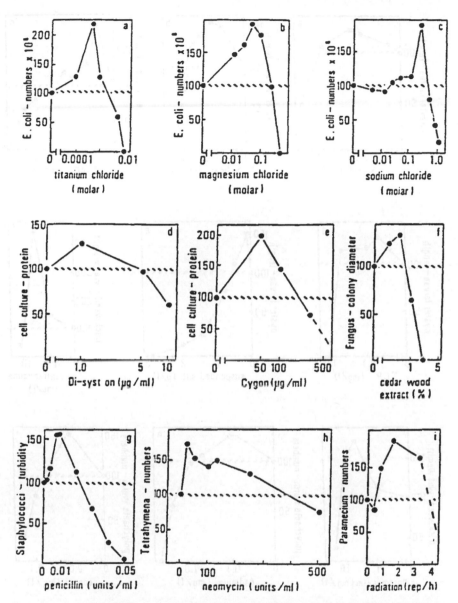

Figure 3.5. The effect of various agents on the growth of organisms and populations drawn from data or figures in the literature. a, b and c, the effect of titanium chloride, magnesium chloride and sodium chloride on the growth of cultures of *Escherichia coli*; d and e, the effect of Di-Syston and Cygon on the growth of cultured mouse liver cells; f, the effect of an extract of western red cedar (*Thuja plicata*) heartwood on the growth of a fungus (*Fomes officinalis*); g, the effect of penicillin on the growth of cultures of *Staphylococcus* (N.C.T.C. No. 6571); h, the effect of neomycin on the growth of cultures of a protozoan (*Tetrahymena gelii*); i, the effect of low levels of beta radiation on the growth of cultures of a protozoan (*Paramecium caudatum*).[20]

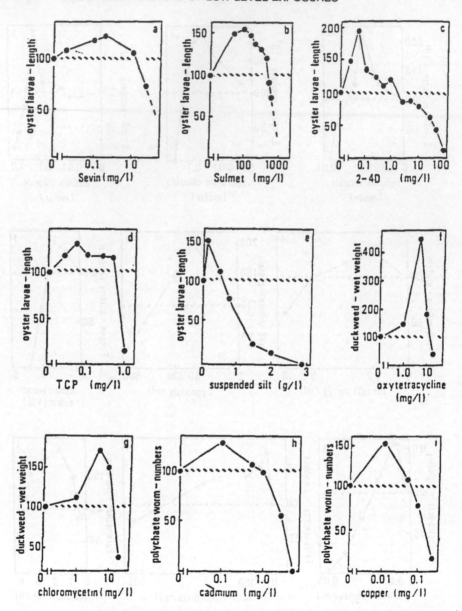

Figure 3.6. The effect of various agents on the growth of organisms and populations drawn from data or figures in the literature. a, b, c, d and e, the effect of Sevin, Sulmet, 2-4D, TCP and suspended silt on the growth of osyter larvae (*Crassostrea virginica*); f and g, the effect of oxytetracycline and chloromycetin on the growth of populations of an angiosperm (*Lemna minor*); h and i, the effect of cadmium and copper on the growth of populations of a polychaete (*Ophryotrocha diadema*).[20]

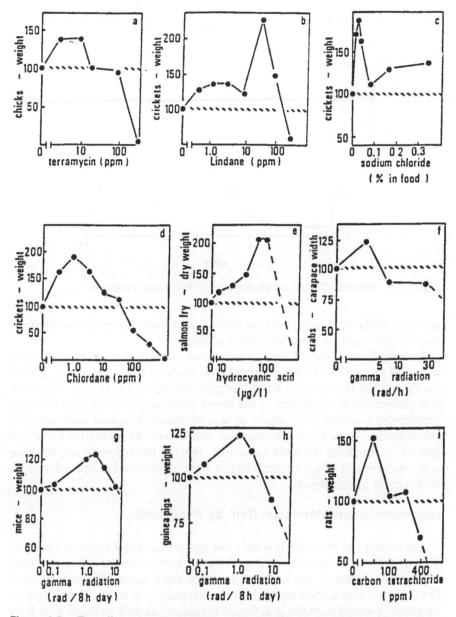

Figure 3.7. The effect of various agents on the growth of organisms and populations drawn from data or figures in the literature. a, the effect of terramycin (oxytetracycline) on the growth of chicks; b and d, the effect of Lindane and Chlordane on the growth of crickets (*Acheta domesticus*); c, the effect of sodium chloride on the growth of crickets (*Acheta domesticus*); e, the effect of hydrocyanic acid on the growth of salmon fry (*Salmo salar*); f, the effect of gamma radiation on the growth of crabs (*Callinectes sapidus*); g and h, the effect of gamma radiation on the growth of mice and guinea pigs; i, the effect of carbon tetrachloride on the growth of rats.[20]

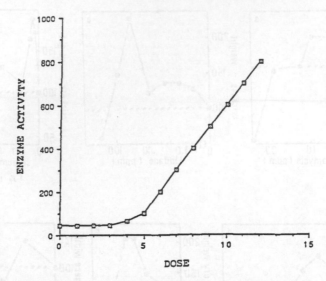

Figure 3.8.   Effect of CCl₄ on serum enzymes 48 hours after exposure.

quently, traditional dose-response studies, therefore, have the possibility of providing insight to hormetic relationships.

Stebbing[20] has argued that one reason that hormesis has not been observed more often is that relative to the study of toxic endpoints, a small proportion of the research has concerned growth of organisms. However, this perspective is challenged by the above discussion that displays hormetic relationships within the context of a traditional threshold dose-response relationship. Thus, hormetic responses could take the form of a B-curve if growth or longevity were the endpoints, but the hockey stick response for toxic endpoints if adaptive mechanisms were operational at lower but overwhelmed at higher doses.

Homeopathy and Hormesis: Guilt by Association

The concept of hormesis is widely but not universally known by toxicologists and environmental scientists. Among those whose training has been biologically oriented, the concept does not seem especially controversial. On the other hand, the concept of hormesis may invoke negative judgment by those involved with the practice of medicine, as well as those who have been involved with reducing exposures to harmful agents via regulatory activities. Why is this the case?

The reasons for these apparent judgments by various groups is not fully known. However, a strong case can be made for the following scenario.

The history of hormesis as noted earlier has its origin with a homeopathic physician Schultz in the 19th century who helped put forth the Arndt-Schutz Law.[6] The linkage with homeopathy was important, since it has lead to both confusion and bias with respect to how the term hormesis is considered.

The practice of medicine known as homeopathy had its origin in the writings of Paracelsus, also the father of modern toxicology. Paracelsus is attributed with creating the "doctrine of similars" when he proclaimed that "what makes a man ill also cures him." He went on to assess this hypothesis by attempting to cure a village of the plague with medicine made with minute amounts of the villagers' own excreta.

It wasn't, however, until the latter part of the 1700s that the modern foundations of homeopathy were established by Hanhemann. The history of homeopathy from the time of Hanhemann to the present has been a controversial and often stormy one.[21]

Hanhemann refined the so-called "doctrine of similars" based on Paracelsus. He described how taking repeated doses of Peruvian Bark (a source of quinine) that caused fever, chills, and other symptoms of malaria was used to treat malaria. He concluded that the reason the bark was beneficial in treating patients with malaria was because it caused symptoms similar to those of the disease it was treating. Hanhemann reported that by optimizing the dose by a series of dilutions, the most effective dose for a patient can be determined. Thus, Hanhemann did several things: (1) reduced the use of large doses of drugs, (2) administered a treatment not to counter the signs and symptoms of a disease but (3) a treatment that would produce the same signs and symptoms if administered in a large enough dose.[21]

Hanhemann's approach for treating patients was not generally accepted by orthodox medicine at the time. So extreme was the response to the claims of Hanhemann and his followers that entire journals were created to oppose homeopathy (e.g., *Anti-Homeopathic Archives, Anti Organan*). Hanhemann also had conflicts with apothecaries.[21] Those individuals favoring Hanhemann claimed the apothecaries were upset because homeopathic practitioners only recommended one medicine at a time and with low doses; thus, it wasn't very profitable for the apothecaries. Hanhemann also argued that the apothecaries often made up the preparations incorrectly or deliberately altered prescriptions so that the treatments wouldn't work. Regardless, one can easily see that much personal and professional conflict could have existed between both groups. In fact, Hanhemann chose to illegally dispense his own medicines, was arrested, convicted and forced to move. However, he was permitted to carry out his trade and dispense medicines by the special permission of the Grand Duke Ferdinand.

Hostility between orthodox medicine and homeopathy was particularly intense in the U.S., the history of which has been documented from both sides of the debate. The significance of this long-standing acrimonious debate was that much of the homeopathic movement which was strong in the 1800s in the U.S. came to a standstill in the mid 1900s.

The hostility of the orthodox medical profession toward homeopathy over turf and economic issues, along with perhaps a genuine disagreement over what is best for patient care and treatment, was compounded by scientific questions about how homeopathic medicine worked. While modern medicine was making great strides in the elucidation of mechanistic understandings of disease causation and treatment, few understandings have been advanced concerning not only how homeopathic remedies work, but if they work at all. The lack of an adequate database with scores of adequately performed double-bind studies and mechanistic insights has been a strong limitation used to weaken the credibility of homeopathy, despite copious anecdotal testimonials. Furthermore, some of the "low" dose treatments used by practitioners are so dilute that it is likely that not one molecule of the original alleged active agent remains in the administered treatment, yet claims are made that it still has clinical effects. The modern mechanistically oriented biomedical field finds it hard to accept claims that are principally anecdotal, with little clinical epidemiological evidence, and no mechanistic understanding.

Given this background of intense historical antipathy along with the weak scientific foundation of homeopathy, the use of minimal optimized doses and the hormesis concept originating with and advocated by a homeopathic physician, it is not unexpected that the hormesis concept is carrying a lot of historical baggage with respect to the medical community, and possibly those toxicologists influenced by education/training programs in medical schools. On the other hand, those more biologically and ecologically oriented individuals would have been less likely to be impacted by the hostility of past interactions between homeopathy and medicine. This may contribute to why hormesis has been a key word in *Biological Abstracts* since 1980, while it is not in *Index Medicus* as of 1994.

The implicit regulatory paradigm that focuses on terms such as no adverse effect levels (NOAEL) not only predisposes investigations to consider not only adverse effects, but bases all dosing schemes on treatments that cause adverse effects in dose-range findings studies. Thus, the system is designed to not only consider adverse effects but it is also a self-fulfilling prophecy based on the type of designs regulatory testing protocols require. Furthermore, given the belief that low dose exposures would show no effect and thus be of little interest to journal readership, the emphasis is placed again on the toxicological effects observed at high exposures. Combine these toxicological perspectives with the belief of environmental advocates that hormesis may be simply a polluter's scheme to justify less stringent environmental regulations, and one can easily see that it is hard to get a fair debate on hormesis.

Future Directions

These perspectives have the capacity to improperly structure the debate over dose-response relationships, especially in the low dose range based on historical antipathies between orthodox medicine and homeopathy. There is confusion over whether hormesis was an integral component of homeopathy, and the concern of public health and regulatory agencies that hormesis means that low dose levels of harmful agents are beneficial. These perspectives have been shown to be essentially historical anachronisms unnecessarily controlling current thought, along with a flawed understanding of the concept of hormesis. Given the powerful responses that the word hormesis evokes in certain influential aspects of the field today, it seems that BELLE should not make hormesis a *cause celebre*, but should restructure the debate in more neutral terms that bring all players to the table. We believe that the focus of BELLE activities should, therefore, be centered on the concept of biological adaptations, how they are affected by dose, and how their expression affects the shape of dose-response curves.

REFERENCES

1. Calabrese, E.J., and K.J. Howe. "Stimulation of Growth of Peppermint (*Mentha piperita*) by Phosfon, a Growth Retardant," *Physiol. Plant.*, 37:163–165 (1976).
2. Cathey, H.M., and N.W. Stuart. "Comparative Plant Growth Regarding Activity of AMO-1618, Phosfon, and CCC," *Bat. Gaz.*, 123(1):51–57 (1961).
3. Kende, H., H. Nineman, and A. Lang. "Inhibition of Gibberellic Acid Biosynthesis in Fusarium Moniliforme by AMO01618 and CCC," *Naurwissenschaften*, 51:599–600 (1963).
4. Knypl, J.S. "Action of (2-Chloroethyl) Trimethyl Ammonium Chloride, 2,4-Dichlorobenzyl Tributyl Phosphonium Chloride, and N-Dimethylaminesuccinamic Acid on I.A.A. and Coumarin Induced Growth of Sunflower Hypocotyl Section," *Acta Soc. Bot. Pol.*, 35(4):611–625 (1967).
5. Picard, C. "Action du CCC et du B995 sur la mise a la fleur d'une plante bisannuell, l'Oenothera biennis (Effect of CCC and B995 on Flowering in *Oenothera biennis*)," *Planta*, 74(3):302–312 (1967).
6. Boxenbaum, H., P.J. Neafsey, and D.J. Founier. "Hormesis, Gompertz Functions, and Risk Assessment," *Drug Metabolism Reviews*, 19(2):195–229 (1988).
7. Smyth, H.F. "Sufficient Challenge," *Fd. Cosmet. Toxicol.*, 5:51–58 (1967).
8. Schulz, H. *Virchow's Archiv.*, 108:423 (1877).
9. Schulz, H. *Pfugers Atch. Ges. Physiol.*, 42:517 (1888).
10. Clark, A.J. *The Mode of Action of Drugs on Cells* (Baltimore, MD: Williams and Wilkins, 1933), pp. 195–197.
11. Clark, A.J. *General Pharmacology* (Berlin: Verlag von Julius Springer, 1937), pp.36–37, 204, 215.
12. Southam, C.M., and J. Ehrlich. "Effect of Extract of Western Red Cedar Heartwood on Certain Wood Decaying Fungi in Culture," *Phytopathology*, 33:517–524 (1943).
13. Brown, R.W. *Composition of Scientific Words* (Washington, DC: Smithsonian Institution Press, 1956), p. 421.
14. Luckey, T.D. in *Proceedings of the First International Conference on the Use of Antibiotics in Agriculture*, Publ. 397, National Academy of Sciences-National Research Council, Washington, DC, 1956, pp. 135–145.

15. Luckey, T.D. in *Recent Progress in Microbiology* (Stockholm: Almquist and Wikell, 1959), pp. 340-349.
16. Luckey, T.D. "Antibiotic Action in Adaptation," *Nature*, 198:263-265 (1963).
17. Luckey, T.D. "Insecticide Hormoligosis," *J. Econ. Entomol.*, 61:7 (1968).
18. Townsend, J.F., and T.D. Luckey. "Hormoligosis in Pharmacology," *J. Am. Med. Assoc.*, 173:44-48 (1960).
19. Luckey, T.D. in "Heavy Metal Toxicity, Safety, and Hormology" (T.D. Luckey, B. Venugopal, and D. Hutcheson, Eds.), Suppl. to Vol. 1 of *Environmental Quality and Safety*, (Stuttgart: Georg Thieme, 1975), pp. 83-103.
20. Stebbing, A.R.D. "Hormesis — The Stimulation of Growth by Low Levels of Inhibitors," *The Science of the Total Environment*, 22:213-234 (1982).
21. Richardson, S. (New York: Homeopathy Harmony Books, 1989), pp. 160.

CHAPTER 4

Commentary on Changing Paradigms: Consistencies and Inconsistencies of Changes in Molecular and Biological Events During Malignant Progression of Transformed Human Cells

George E. Milo, The Center for Molecular Environmental Health, Department of Medical Biochemistry and The Comprehensive Cancer Center, The Ohio State University, Columbus, Ohio

INTRODUCTION

The rationale for using human cells in culture to evaluate the action of xenobiotics is that many different cell types have the capacity to activate xenobiotics. These metabolites may then interact with various parts of the nuclear machinery. Specific modification of the molecular machinery that leads to a molecular mutation can induce a sequence of phenotypic changes that can lead to the biological expression of a transformed cell. There is relative agreement among many investigators that these progressive sequential changes occur following the formation of the initiated cell, and are consistent with a multistage process. In the progression of an initiated cell toward a malignant stage, the timing between the stages of the formation of the initiated cell, expression of anchorage independent growth (AIG) stage, and the neoplastic stage appear to be an inconsistent variable. These molecular changes result in cellular changes that can be identified during the progressive development of a tumorigenic phenotype.

The inconsistency in timing of the length of the individual stages in progression may be related to the culture environment and/or type of treated cell. Another variable appears to be the observation that there is a lack of a direct correlation between the biological endpoint mutagenesis and carcinogenesis, when different dosage levels of the activated xenobiotic were used to initiate the process. McCormick and Maher[1] and Silinskas et al.[2] indicated that there are direct consistent relationships between dosages levels used to treat cells and to elicit either a mutagenic or carcinogenic response. Inconsistencies do exist over a wide dosage range of treatment and are explainable. Transformation (AIG or tumorigenicity) is a biological endpoint that requires that compensatory cell proliferation of the initiated phenotype occur selectively prior to the treated cells' progress toward a more malignant phenotype, not a general mitogenic response as assumed. One also may have selective cell proliferation or accumulation of initiated cells without cell death (selective mitogen for the initiated cell or inhibition of apoptosis). Cytotoxicity, or cell death, dictates removal of the treated cell from the proliferating cell population and does not permit these cells to become part of the transformed cell population. Transformation and cell death are opposing biological phenomena. It is conceivable that these two different biological processes by themselves may constitute responses to toxic dosages in areas of the treatment curves where there may be direct inverse correlations;[2] however, over broad toxic dosage ranges these direct correlations do not appear to exist. The weakness in examining relationships between toxic endpoints and transformation is the overinterpretation of the results over a limited range of exposure to the different toxic concentrations of either the mutagen or carcinogen. Cells treated at different toxic dosages may exhibit either contact inhibition, cellular senescence, or program cell death as a result of the toxic treatment. More often than not, there have been repeated attempts presented in the literature to justify the use of cell lines that exhibit "infinite life spans" to undertake these studies to negate or reduce the noncorrelative toxic responses of contact inhibition or cellular senescence. For example, cells that exhibit "immortality" have an abbreviated/ altered G_1 quiescent stage and/or G_0 stage. An example of such a nonmalignant transformed cell line that exhibits a loss of senescence, G_1 quiescence is to use SV40 transformed immortalized cells. Immortalization, a word coined in the late 1950s or 1960s, was used to describe nonexpressing SV-40 transformed cells. The most notable of these cells were derived from the SV-40 transformed WI-38 and WI-26 embryonic human lung cells. The SV-40 persistent transformed counterparts were labeled 13–2RA and VA-4 cell lines. There are many examples in the literature of persistent attempts to establish other immortalized cell lines on a consistent basis. Most of these examples represent a lack of understanding of how SV-40 transforms human cells and alters the expression of specific parts of the cell cycle. In any event, there are certain features associated with using cells with a finite life span that can lead to alternate interpretations. For example, one such experimentalist[1] stated that when one cell survives treatment at a specific cytotoxic dose and proliferates to 10^6 cells, that this represents an increase of 20 population doublings (PDL). No definition of cytotoxicity was presented in the manuscript. Mathematically, if we use the mathematical model developed by Leonard Hayflick in the early 1970s to define population doublings in human cells:[3]

$$n = 3 \cdot 32$$
$$(\log N - \log X_0) =$$
$$n = (\log 10^6) - (\log 10^0)$$
$$n = (6) - (1)$$
$$n = 5 \text{ PDL, not } 20.$$

The liberty of arbitrarily assigning a PDL of 20 is not appropriate.[1]

If we assume 5-7 PDL is correct and these numbers of PDL are again required before a second "acquired trait" is introduced, then the PDL will be 10-14 PDLs, not 39 PDLs, as determined by applying the above rigorous mathematical equation, since the authors[2] state that expanding the population from one cell to 10^6 cells yields another 20 PDL. To arbitrarily assign a numerical value to PDL does not establish a rigorous scientific rationale for determining when additional characteristics associated with initiated cells in different stages of progression will appear. It is generally considered by the people in aging research that a cell derived from embryonic origin exhibits ca. PDL ~ 60 before the culture of human cells enters Phase III senescence. Cells from neonatal human tissue generally exhibit PDL ca. 35 ± 3 PDL before entering the post mitotic stage. In cell populations in culture derived from *older* tissue sources, the PDL drops dramatically with increasing age of the donor. Data developed in vitro suggest that carcinogen-induced phenotypic changes in cells develop characteristics similar to some individual cell populations prepared from human tumors.[4] These tumor-associated characteristics are either the expression of a phenotype that expresses a cell surface tumor associated antigen, AIG, cellular invasiveness, tumorigenicity, or expression of a malignant metastatic potential in a surrogate host.

It is generally agreed upon in the scientific community that the activated forms of environmental xenobiotics results in adducts to specific bases in the genomic DNA. These observations should dictate that a comparison of the biological endpoints, cytotoxicity, transformation and mutations with adduct modification should exhibit linear relationships if there are direct causally related associations between the biological events and the DNA adduction at different treatment dosage levels (and many assumptions being made). The nature of dissimilar biological endpoints would, by themselves, suggest that each of the endpoints and adduct formation would exhibit indirect correlations.

The lack of a strong linear relationship between different biological endpoints and the extent of modification of specific bases in the genomic DNA could complicate the mechanistic interpretation of both the carcinogenesis and mutagenesis assays and the biological relationships, if such exist, between mutagenesis, as measured by a lesion in the *hprt* gene, and transformation.[5] The association between major and minor specific-base modification, and either cell transformation or mutagenesis, at best, serves as an approximation of a general relationship between a biological endpoint and modification of specific target molecules. We assume the "mutation" assays only detect mutations and not epigenetic changes, and that the transformation experiments pick up only mutational events and not epigenetic events.

In other mammalian systems[6], incongruities have been observed when mutagenesis is correlated with transformation over a range of concentrations of either mutagens or carcinogens.[7] For example, Elmore et al.[8] showed that increased mutation rates in carcinogen-treated human fibroblasts did not exhibit a strong correlation with induced transformation frequencies of human fibroblasts over a range of dosages from nontoxic to toxic.

Many compounds present in the environment may require activation of the parent compound in order to present the activated metabolite to the reactive cellular substituent group. In order to produce such activated metabolites, S9 mixes have frequently been used. However, anomalies have been observed when the specific activated metabolites such as BPDE-I have been compared to the metabolites of S9 mix-activated B(a)P and parent B(a)P treated fibroblasts. These data apparently suggest that the P450 complexes of each of these systems form a different spectrum of metabolites. Recently, we [9,10] reported that freshly isolated human foreskin cells in vitro can metabolize B(a)P into a transformation active metabolite without adding an S9 mix. Although an S9 mix used in human cell toxicology studies biotransformed the B(a)P, the quantitative and qualitative distribution of the metabolites were different when compared to the endogenous metabolites produced by cellular metabolism of B(a)P.[9] This observation may account for the greater amount of adduct formation when the cells metabolized the B(a)P without an exogenous S9 mix. Another observation we made was that exogenously-supplied S9 mix to the cells in the presence of B(a)P produced a two- to fourfold increase in tetrols found outside the cell, compared to the intracellular amount.[15,16] These distributions of extracellular and intracellular metabolites of intact cell metabolism of B(a)P were lower in amount, and the extracellular versus intracellular amounts was not as exaggerated as the S9 mix metabolism. The transformation frequencies and adduct modifications were similar when both approaches were used. These inconsistencies between S9 and intact human cells in metabolism led to a misunderstanding of the role of metabolism and formation of metabolites in the mutagenic and carcinogenesis process.

Modification of DNA by a direct-acting carcinogen[11,12] leads to the formation of initiated cells that exhibit anchorage-independent growth but do not exhibit the feature of uneven nonquantitative distribution of metabolites as observed for B(a)P activation systems. It is our opinion that the use of exogenous activation systems to metabolize carcinogens can produce ambiguous and erroneous results. The extent of metabolism by exogenously supplied S9 mix definitely affects the distribution of intracellular, nonbound metabolites. As the concentrations of the activated compounds increased, the extent of modification of specific dG molecules also proportionally increased; however, it was interesting to note that we did not see a corresponding direct increase in the expression of transformation.[9,10] The increase in DNA modification did not directly correlate with a corresponding direct increase in either toxicity or transformation over a broad concentration range. These observations are especially appropriate when cells were treated at low nontoxic dosage levels of the compounds.

In summary, the first change in the expression of a normal cell to a carcinogen-transformed phenotype following treatment with a carcinogen may be an altered cellular morphology. The presence of morphologically altered cells depends on the cell type and the carcinogen used to induce the initiation stage. Moreover, the altered morphology usually occurs only transiently among 2 and 10 PDLs following treatment. At various times after treatment and following passage in culture fibroblast-treated cell populations exhibit AIG. These observations suggest that the treated cells progress toward a malignant phenotype with passage in culture. Treated cell populations that subsequently exhibit AIG, followed by the expression of cellular invasiveness on chick embryonic skin, will then form either an intracranial or subcutaneous localized tumor 0.8 to 1.2 cm in size in a surrogate host. These localized tumors may grow progressively or regress.

These stages of progressive development in the carcinogen-transformed human cells are consistent with the concept of multistage development of cancer cells [but do not necessarily indicate that these phenotypic changes occur in a solid tissue in vivo], since mixed tumor phenotypes isolated from spontaneous occurring human tumors also exhibit many of the same stages of progression and phenotypic diversity, such as AIG, specific cell surface tumor associated antigens, and limited tumor growth in nude mice. Moreover, the progressive development from a subcutaneous nodule to a progressively growing tumor in the surrogate host is not routine. There are many instances where carcinogen-transformed epithelial or fibroblast cells and cells from either human sarcoma or carcinomas regress. This brief review discusses possible correlative and causal relationships between carcinogen-induced molecular events leading to transformation and multistage progression in carcinogen-treated human cells. Where appropriate, consistencies and inconsistencies in each of the stages of tumor progression of either carcinogen treated cells will be compared with cell populations prepared from tumors from patients.

DISCUSSION OF INCONSISTENCIES

Figure 4.1 graphically represents the presently accepted interpretation of linear malignant progression.[13]

An interesting feature of this linear interpretation is that the stage identified as the promotion stage in human cell carcinogenesis is for all practical purposes at this time, a silent stage. However, if we accept the premise that there is a clonal expansion of initiated cells following treatment with a

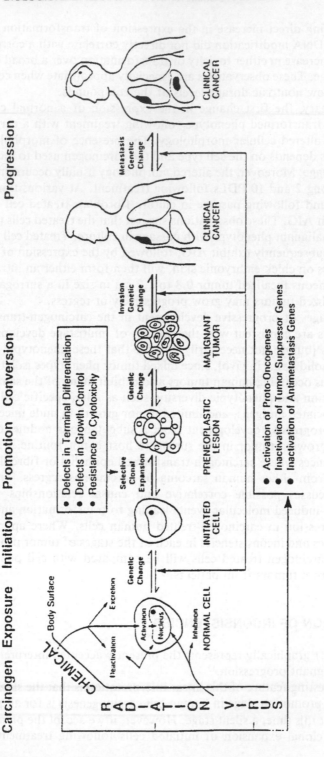

Figure 4.1. This presentation is an attempt to define, in a linear manner, multistage carcinogenesis. The stages presented in this figure represent the stages in progression from a normal cell exposed to a carcinogen through a stage identified as chemical-induced cancer.[13]

carcinogen, we must accept by definition that the promotion stage is operational. However, the animal promoters, TPA, anthralin, benzoyl peroxide, and messarine are, for the most part, antiproliferative agents[14] when applied to initiated human cells. In addition, this linear scheme cannot accommodate the concept of genotypic and phenotypic diversity. The model presented in Figure 4.2 attempts to deal with this issue. For example, the phenotype JC1 (Figure 4.2) has a molecular mutation in a *ras* gene, p53 and NM23 + gene, whereas JC2 has a mutation in *ras*, p53 but not a molecular mutation in NM23. Here we have an example of genotypic diversity and phenotypic homogeneity, i.e., both phenotypes exhibit the same degree of tumorigenic vigor. One must recognize the reality that the NM23 gene, a metastatic suppressor gene, may not play a direct role in metastasis.

To address these divergent phenomena we must first examine the individual stages from metabolic activation through malignant progression. One compound that can be used to examine the activation stage is benzo(a)pyrene (Figure 4.3).[15]

Other compounds as they impact on our understanding of the development of stages in the carcinogenic process will be introduced into the text of

Phenotypic / Genotypic Diversity

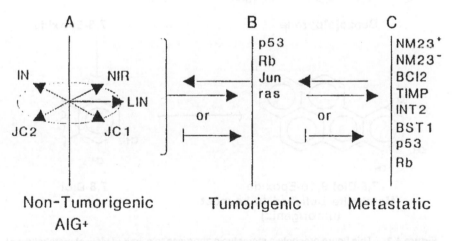

Figure 4.2.   The designation A; B; C represents a nonlinear dimension of time associated with tumor progression. In this model we recognize phenotypic and genotypic diversity. The phenotypes are designated IN, NIR, JC1, JC2 and LIN at stage A. We have identified stage A as a nontumorigenic anchorage independent growth stage (AIG +). The acronyms in either stage B or stage C are designated specifically with the stage of either tumorigenicity or metastasis. The surrogate host used to evaluate these stages is the nu/nu mouse (see Reference 18). The acronym p53 refers to the p53 tumorigenic suppressor gene. Rb refers to the retinoblastoma gene. *Jun* and *ras* are oncogenes. NM23 refers to a metastatic suppressor gene. BCL2, TIMP, INT2 and BST1 are genes associated with either the tumorigenic or metastatic stage.[18]

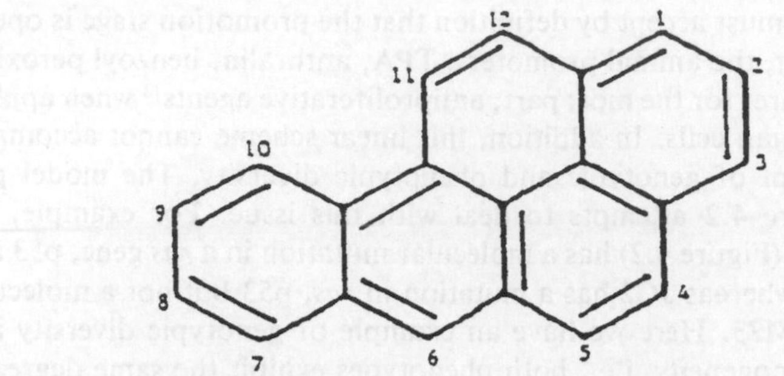

Benzo[a]pyrene

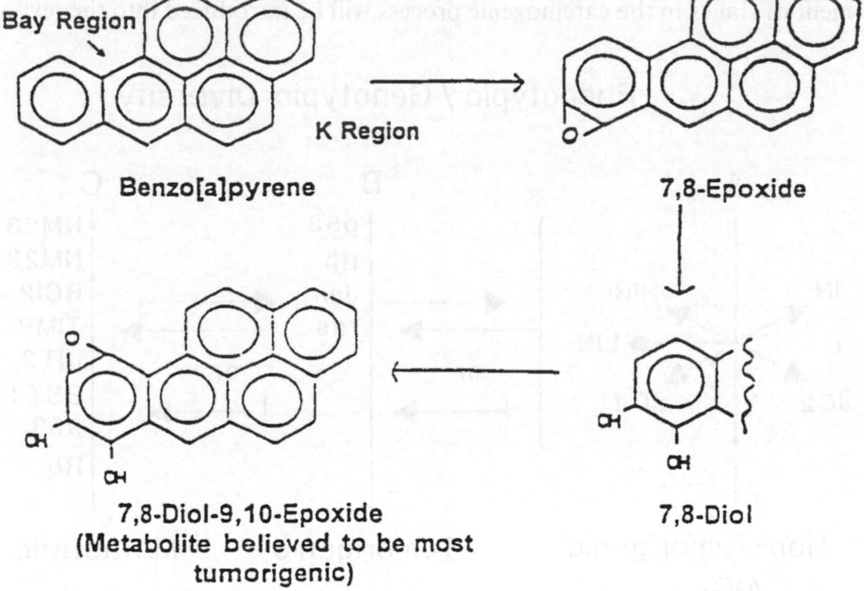

Figure 4.3. This figure presents a stereotypic approach to a limited view of metabolism of B[a]P to its activated metabolites.

the discussion. There are two isomers of the activated B(a)P that are immediately formed by the induced P450 complex; the anti-isomer of 7,8 diol-9,10-epoxide I and II (Figure 4.3). There are also syn isomers formed that are part of the profile of metabolites. The one we will focus on is the anti (I) (+) enantiomer. This isomer forms an adduct with dG nucleoside in excess of 95% of the detectable BPDE-I-N7dG adducts. The quantitative and

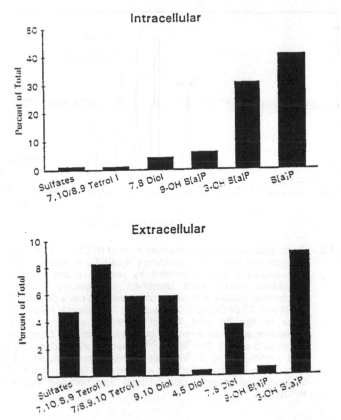

Figure 4.4. This histogram presents the intracellular and extracellular distribution of B[a]P metabolites as a percentage of the total amount of metabolites.

qualitative intracellular and extracellular distribution of the organic soluble and water soluble metabolites are decidedly different[15] (Figure 4.4).

When the major metabolite BPDE-I anti form was used to treat cells, the formation of 7S and 7R adduct (BPDE-I-N7dG) was observed (Figure 4.5).[16,17] When comparing the internal distribution of the 7,8-dihydro B(a)P-7,8-diol metabolite[15] and the BPDE-I-N7dG adduct levels, a substantial amount of the cpoxidc form in the cell was found not to react with dG (Figure 4.6). This same relationship holds at a 1/3-fold increase or decrease in dosage levels of BPDE-I treatment. When the concentration of the 7,8-dihydriol diol was elevated or decreased by 2.0 or reduced by 0.5, the increase in BPDE-I-N7dG adducts is not proportionally decreased or increased. When we compare specific adduct modification of BPDE-I-N7dG[16,17] with either methylazoxymethanol or 1-nitrosopyrene, modification of dG, there was a relative equal amount of adducts formed by each carcinogen (Table 4.1). However, there was an inconsistent variation in the incidence of colony formation (Table 4.1) at equi-levels of adduction modi-

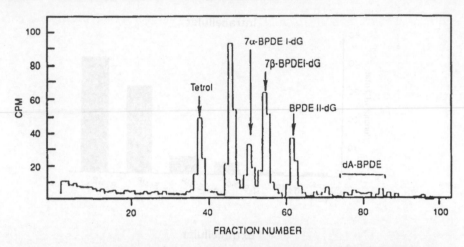

Figure 4.5. A representative h.p.l.c. chromatogram of BPDE-I-DNA deoxyribonucleo-side adducts was obtained by reacting [³H]BPDE-I with calf thymus DNA in vitro. The adduct(s) were separated by reverse phase chromatography. The individual peaks were identified by relative elution time and/or co-chromatography with known standards. 7R-*trans* dG and 7S-*trans* dG refer to 7R and 7S BPDE-I-dG. 7R-*trans* dA, 7S-*trans* dA, 7S *cis* dA and 7R *cis* dA refer to the 7R and 7S BPDE-I:*trans* and *cis*-N6-dA. Insert: hplc chromatogram of the methanol eluting fraction from Sephadex LH-20 for the main dpNp adduct. The ³H and ³²P-labeled dpNp adduct was isolated with tlc and hydrolyzed to a ³H-labeled deoxyribonucleoside adduct with acid phosphate. The ³H-labeled was isolated from Sephadex LH-20 with methanol and analyzed on hplc. One ³H peak was obtained and identified as 7R BPDE-I-dG.[17]

fication and from treatment to treatment, and over a defined toxic dosage range. In fact, if we eliminated the formation of a transformed phenotype in the presence of benzamide, adduct modification was approximately at the same level in the cells that are transformed.

When we examined the relationship between concentration (dosage μg/mL), expression of AIG and adduct modification with B(a)P, BPDE-I and 1NP, similar discrepancies (Table 4.2) were observed.[18,19] As the concentration of BPDE-I was increased to substantive toxic levels where 75% of the cells were killed, the number of adducts determined/10^6 nucleotides plateau whereas the number of transformed phenotypes decreased from a high of 121 ± 6.2 to 12.0 ± 3.2. When we inhibited the formation of the AIG+ phenotype in soft agar by the addition of 1mM nontoxic level of benzamide, the interesting feature was that DNA adduction was consistently lower but not statistically significantly lower.

These discrepancies appear to be most pronounced when cytotoxicity and the frequency of expression of AIG are compared. As the concentration of BPDE-I increased from 60% to 70% cytotoxicity, the adduct level was relatively unchanged, but the expression of AIG dropped from a high of 121 colonies per 10^5 treated cells to 12 colonies (Table 4.2). A similar observa-

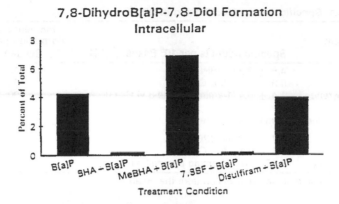

7,8-DihydroB[a]P-7,8-Diol Formation
Intracellular

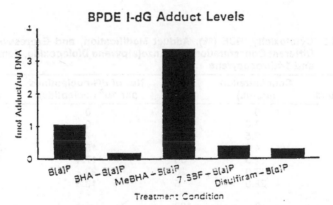

BPDE I-dG Adduct Levels

Figure 4.6. Data presented here represent what happens to benzo[a]pyrene when it is metabolized to its intracellular dihydrodiol, an indirect measurement of the production of the B[a]P epoxide. BHA (butylated hydroxyamisole) MeBHA (methylated form of BHA), 7,8BF (7,8 benzo flavone), and disulfiram are reported intervening agents in the metabolism of B[a]P. MeBHA does not interfere in the metabolism of B[a]P to its dihydrodiol epoxide but eliminates the expression of a AIG⁺ phenotype.

tion was also made when cells were treated with 1-NOP and MAMA (data not shown). At dosage levels where no cytotoxicity was observed, the occurrence of both adducts and transformants were observed. Apparently the biological expression, cell death, is for the most part unrelated and non-correlated with metabolism, adduct modification, and cellular transformation at low dosage levels. A different question then was asked! If the cells are treated at different concentrations and at different stages of the cell cycle, will different transformation frequencies be observed? Table 4.3 presents data that suggest that there was a discrepancy in the expression of the AIG phenotype when cells were treated in either S_1, or early S compared to late S (S_3) or the G_1 late stage of the cell cycle. There was a >2 log increase in the expression of AIG when cells were treated in early S as

Table 4.1. Specific Adduct Modification of BPDE-1

Treatment Agar	Specific Adducts per 10^6 Bases $\pm$ S.D.	Incidence of Colony Formation per 10^5 Cells Seeded in Soft
BPDE-I	4.5 $\pm$ 7.2 [7-βBPDE-I G]	50–80
MAMA	5.3 $\pm$ 3.7 [Me]$_6$ G]	400–800
1-Nitropyrene	1.4 $\pm$ 0.4 [1-amino pyrene-8 yl C$_8$ G]	100–200
In the presence of 1 mM benzamide		
BPDE-I	3.7 $\pm$ 2.8 [7-βBPDE-I G]	–3–
MAMA	4.8 $\pm$ 3.1 [Me]$_6$ G]	–5–
1-Nitropyrene	1.46 $\pm$ 0.4 [1-amino pyrene-8 yl C$_8$ G]	–2–

S.D.I = 1 sigma standard deviation from the mean.

Table 4.2. Cytotoxicity, RCE (%), Adduct Modification, and Expression of AIG at Different Concentrations of Benzo(a)pyrene Diolepoxide, Benzo(a)pyrene, and 1-Nitrosopyrene

Compound	Concentration (μg/mL)	RCE (%)	No. of dG-nucleotides per 10^6 nucleotides	Frequency AIG $\pm$ S.D.
BPDE-I	0	100	0	0
	1.0	100	0	0
	2.5	100	7.7	95.0 $\pm$ 8.7
	4.6	50	15.3	121.0 $\pm$ 6.2
	7.5	35	17.4	70.0 $\pm$ 12.0
	17.0	20	23.2	12.0 $\pm$ 3.2
	25.0	6	18.3	3.0 $\pm$ 2.0
B(a)P	0	100	—	0
	0.1	100	12.4	29.7 $\pm$ 14.5
	2.5	100	18.6	16.0 $\pm$ 12.0
	10.0	100	13.4	103.0 $\pm$ 41.0
	25.0	100	11.7	101.0 $\pm$ 2.8
1-NOP	0	100	0	0
	5.0	100	0.2	0
	7.5	100	0.7	5.0 $\pm$ 3.0
	10.0	92	7.0	6.0 $\pm$ 2.0
	18.0	80	11.0	15.0 $\pm$ 4.0
	30.0	60	24.0	23.0 $\pm$ 2.0
	95.0	50	51.0	41.0 $\pm$ 7.0
	105.0	21	60.0	10.0 $\pm$ 12.0

$\pm$ S.D. (see Table 4.1).

Table 4.3. Number of Colonies Formed in Soft Agar (0.33%) per 10^5 Seeded Cells

	S1	S2	S3	G1 Release
EXP 1	574 $\pm$ 31.4	130 $\pm$ 8.0	5 $\pm$ 2.3	2 $\pm$ 0.7
EXP 2	382 $\pm$ 17.6	75 $\pm$ 3.2	1 $\pm$ 0.8	0

$\pm$ S.D. (see Table 4.1).

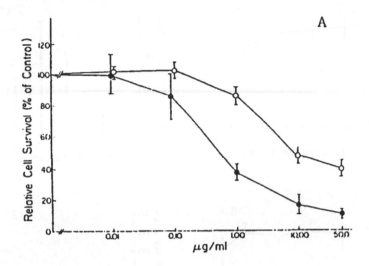

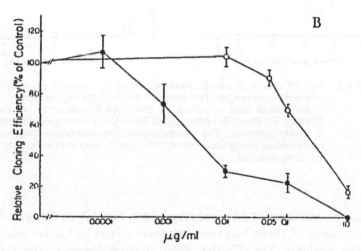

Figure 4.7.   Response of the cells to chemical using trypan blue exclusion. Panel A presents data as a function of changes in concentration of T-2 (-p-) and T2-tetraol or 4 log ranges of concentration. In Panel B we measured the cytotoxic effect of the drugs at different 4 log dosage. Levels are calculated using the relative cloning efficiency of the human cells. The vertical bars represent 1 S.D. from the mean.

opposed to late G_1 treatment. However, the extent of BPDE-I modification of the total genomic DNA was comparably similar at each stage. [17] To discern whether the methodology used to determine either transformation or cytotoxicity is a valid approach, we examined two approaches to determine cytotoxicity.

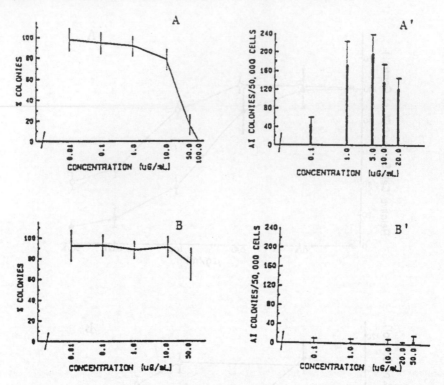

Figure 4.8. The left column of panels, Panel A through D, represent cytotoxicity curves where different ranges of the compounds inhibit the relative cloning efficiency of the human cells in culture, as described in Reference 10. Panel A is Aflatoxin B1; Panel B is Aflatoxin B2; Panel C is 1-Napthylamine; and Panel D is 2-Napthylamine. The corresponding transformation frequencies are presented as histograms, Panel A' through D,' over similar noncytotoxic and cytotoxic dosages.

Mycotoxins T-2 and T-2 tetraol (Figure 4.7, Panel A and B) were selected as compounds of choice[20] to study the toxic effects on human cells by two different methods. The effective 50% cytotoxic dosage of the T-2 mycotoxin when measured by trypan blue exclusion was ca. 0.90 μg/mL (Figure 4.7), but when measured by its effect on cell proliferation it was ca. 0.005 μg/mL. A similar response pattern was observed for T-2 tetraol (Figure 4.7, Panel B). The shapes and slopes of the toxicity curves were quite similar; however, the observed individual quantitative toxic values were quite different. To illustrate variations in cytotoxicity patterns of other related compounds, the cytotoxicity of Aflatoxin B1 (Figure 4.8A) and Aflatoxin B2 (Figure 4.8B) were compared. This approach was extended to measuring the transformation frequency (AFB1 — Figure 4.8A and AFB2 — Figure 4.8B) at the different concentrations of the chemical.[10,21]

From the data presented in Figure 4.8A, at concentrations of Aflatoxin

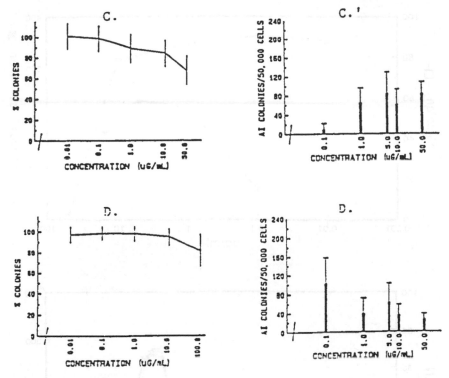

Figure 4.8. Continued.

B1 from 0.01 to 0.1 μg/mL, there is little or no toxicity associated with this chemical, although Aflatoxin B1 treatment of the cells at these concentrations quite decidedly transformed the cells, Panel A'. In contrast to these results, Aflatoxin B2 (Panel B) exhibited a relative minor toxic effect over a much broader range of dosages, and only a minor transformation response was observed even at higher concentrations, (Panel B'). In Figure 4.8C' and 4.8D', significant numbers of transformants were observed when cells were treated with either 1-naphthylamine or 2-naphthylamine over a 1 to 2 log difference in concentration. The data presented in Figure 4.9, Panel A, present a comparison of cytotoxicity (-□-) following BPDE-I (+) treatment with the number of AIG colonies formed in soft agar, (-■-).[22] Again, these data suggest that there is not a linear relationship between the expression of transformation, AIG, and cytotoxicity, as mentioned previously with B(a)P metabolites or other nonbulky carcinogens. We compared the profile of transformation response with cytotoxicity (Figure 4.9). Over a full treatment range (Figure 4.9, Panel A) when cells are treated at different dosages (Figure 4.9, Panel B) we did not observe a linear response.

The treatment regimen is illustrated in Figure 4.10;[23,24] the carcinogen,

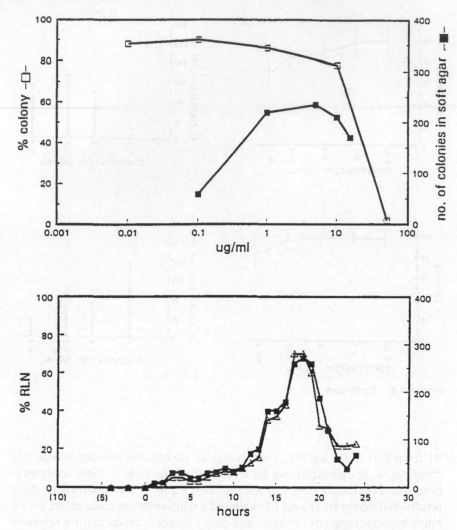

Figure 4.9. The upper panel represents a cytoxicity profile of BPDE-I treated cells (-□-) compared with the transformational frequency (AIG⁺) phenotype, (No. of colonies ≥60μ in size formed in soft agar). The bottom panel presents an experimental radiolabeling curve similar to that presented in Figure 4.10. See legend of Figure 4.10 for details of treatment with BPDE-I. BPDE-I was left on the cells for 30 min in early S. Note the formation of AIG⁺ colonies at nontoxic levels of BPDE-I, (1.0 μg/mL to 10⁻² μg/mL).

BPDE-I (+) was administered for 60–90 minutes and then removed. Treatment of the human fibroblast cultures, in early S, with either a nontoxic transforming dosage level or a toxic transforming dosage, resulted in a molecular mutation in the 12th/13 coding region of the Ha-*ras* protooncogene (Figure 4.11).

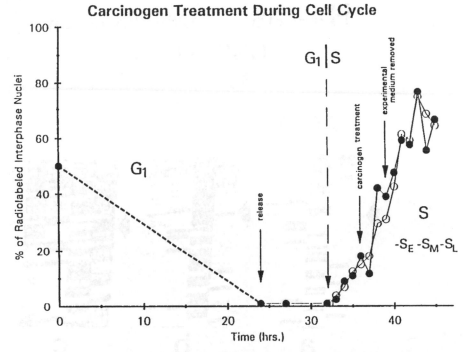

Figure 4.10. The human fibroblast cultures are blocked in G_1 part of the cell cycle. The radiolabeling index (RLI) of these randomly proliferating cells in logarithmic growth is 43 ± 2.0%. In the blocking medium the RLI dropped to <0.1%. The cells are released by adding back the deleted amino acids, glutamine and arginine. S phase entry occurs 10 hr later. This designation is described as G1/S period. S_E is designated as early S, 0–3 hr after S phase entry. S_M is designated as middle S, 3–6 hr after S phase entry. S_L is designated as late S period, 6–10 hr after S phase entry. The time when carcinogen treatment takes place and when the experimental treatment medium is removed generally occurs over a definite time intervals in S_E.

CONCLUSIONS: FACT OR FICTION!

We[10,19,23,24] and others[7,8,11,25] have shown that environmental insults, either physical[24] or chemical, can induce cellular transformation. However, it must be recognized that biological endpoints of transformation when compared with changes in molecular adduction may be complicated by the structure of the cell. For example, activated chemicals may form adducts in DNA contained in mitochondria and to amino groups in other proteins; to other nucleic acids and to cellular glycoproteins. These other classes of biologically active substances play substantial roles in cell viability, cellular proliferation, cell death, and cell transformation. Alteration in genetic information may ultimately lead to the expression of abnormal pathologies that result in the formation of soft tissue tumors. In these soft tissue tumors, breast cancer, carcinomas, sarcomas, etc., the tumorigenic stage

Ha-ras   12th/13th codon
GGC → GTC
Gly   Val

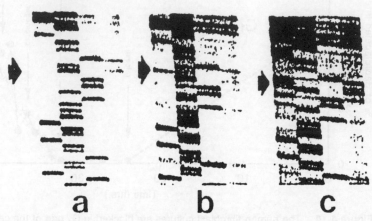

Figure 4.11. This figure presents data on three autoradiograms comparing GGC sequences of the Ha-*ras* 12th codon region. DNA templates for the sequencing were prepared by conventional PCR analyses. Asymmetrical PCR was carried out using amplified DNA with specific primers for the codon region. Sequencing reactions were labeled with [[^{35}S]thio]dATP and one of the primers. The products were resolved by sequencing gels. Panel a represents a normal human fibroblast cell population. Panels b and c represent BPDE-I treated AIG$^+$ cells.

may involve changes in several different chromosomes, such as 51, 17P, or 18q. Colorectal tumors have demonstrated the rearrangements of these aforementioned chromosomes and, in addition, two additional genes, p53 and DCC.[26,27] In melanoma, soft tissue tumors rearrangements involving chromosomes 1, 6, and 7 have been implicated.[27] However, these rearrangements have involved a consistently low percentage of the total number of cells, 26% to 30% in a tumor.[27] The molecular mutation in colorectal tumors, activated p53 tumor suppressor gene, was found in approximately 26% of the tumors.[26] In the melanoma tumors, one of five tumors have an activated *ras* gene;[28] it would appear as though *ras* activation may not be necessary for transformation progression. We have found that the presence of activated H-*ras* in the 12th coding region does not predetermine that the molecular mutation to a causal role in the expression of a tumorigenic phenotype; however, there is definitely a correlative role. At the present time we cannot critically define the role of this lesion in H-*ras*.

The induction of the initial events, e.g., DNA modification, requires that the genomic DNA occupy a specific conformational configuration[17,23] in order for the activated form of the carcinogen to modify specific sites in the gene areas. Conformational change in the dynamic stage of chromatin-DNA can result in no modification of specific base sequences in active gene areas. Our data also suggest that there does not appear to be an absolute requirement for a cytotoxic level of carcinogen to be present to observe a molecular mutation and/or altered biological endpoint. While linearity may appear to exist in examining relationships between transformation and cytotoxic dosage levels of compounds over a limited dosage range, we have observed many different patterns of inconsistent nonlinear responses to the treatment with the activated carcinogens. The subsequent sequential progressive change in function of these transformed cells is consistent with a multistage process. Phenotypic and genotypic diversity in cell populations prepared from human tumors is inconsistent with the process of progression being a linear process. These observations are consistent with the viewpoint that other factors can alter the temporal progression of malignant DNA when transfected cells are not immediately transformed into malignant phenotypes. If we accept the fact that multiple mutagenic hits in specific hot spots of either oncogenes of suppressor genes play an important role in malignant progression, then we must also accept the fact that nonlinear biological response argues for phenotypic diversity and tumorigenic heterogeneity. The lack of strong linear relationships between different biological endpoints complicates the mechanistic interpretation of both carcinogenic and mutagenic endpoints. Reduced to a molecular level, the association between minor and major specific-base modification leading to either an expression of either cellular transformation or mutagenesis may, at best, serve as an approximation of a suitable endpoint.

It is my considered opinion that the paradigms proffered by many investigators to explain these carcinogenic events is/are at best appropriately designed to aid us in interpreting the results. However, by themselves these overinterpretations have not shed a great deal of light on the complexity of the biological process of environmental carcinogenesis induced cellular transformation.

It is appropriate to recognize that from the moment of exposure DNA damage and non-DNA damage (altered gene expression) to the stage of a biological endpoint, that there is a complex dynamic series of events taking place. Our naive attempt to assign complex biological endpoints as the consequence of individual molecular changes at the gene level is contributing to an oversimplification of what is actually taking place inside the cell.

ACKNOWLEDGMENT

The data presented in this manuscript represent the culmination of research by many different investigators and students, either in residence or collaborators with George E. Milo. For the most part, each participant is recognized as a co-author to Dr. Milo on each published manuscript. In particular, I would like to acknowledge the most recent finding of Dr. Hakjoo Lee for her PCR studies and epigenetic modulation data; Dr. Jucheng Chen for his advice and findings in molecular pathology of the stages in progression of the squamous cell carcinoma; Dr. Dawei Li for his persistence and contributions in developing substraction libraries of the transformed cells in different stages of malignant progression. I would like also to acknowledge the professional contributions of Dr. Charles Shuler in discussions with me regarding the concept of phenotypic plasticity and phenotypic diversity.

I would like to express my appreciation to Ms. Inge Noyes for her technical assistance, and Ms. Martha Leming for her administrative support.

I would like to convey my appreciation to NIEHS R01 ES04623 (GEM), NCI R01 CA25907 (GEM,CS) and EPA R815605 (GEM) for their support of these projects that resulted in the author's writing this manuscript.

REFERENCES

1. McCormick, J.J., and V.M. Maher. "Towards an Understanding of the Malignant Transformation of Diploid Human Fibroblasts," *Mutation Res.*, 199:271 (1988).
2. Silinskas, K.C., S.A. Kateley, J.E. Tower, V.M. Maher, and J.J. McCormick. "Induction of Anchorage Independent Growth in Human Fibroblasts by Propane Sultone," *Cancer Res.*, 41:1620 (1981).
3. Hayflick, L., "Subculturing Human Diploid Fibroblast Cultures," in *Tissue Culture: Methods and Applications*, P.F. Kruse and Patterson, M.K., Eds., (New York: Academic Press, Inc., 1973), Chap. 3.
4. Kumari, H.L., C. Shuler, D.G. Mannix, and G.E. Milo. "HNF Transformation with Chondrosarcoma DNA Results in the Development of a Sarcoma Cell Surface-Associated Epitope," *Exptl. and Molec. Pathol.*, 53:167 (1990).
5. Barrett, J.C., and P.O.P. T'so. "Relationship Between Somatic Mutation and Neoplastic Transformation," *Proc. Natl. Acad. Sci.*, 75:3297 (1978).
6. Huberman, E., R. Mager, and L. Sacho. "Mutagenesis and Transformation of Normal Cells by Carcinogens," *Nature*, 264:360 (1971).
7. Barrett, J.C., T. Tsutsui, and P.O.P. T'so. "Neoplastic Transformation Induced by a Direct Perturbation of DNA," *Nature*, 274:229 (1978).
8. Elmore, F., T. Kakumaga, and J.C. Barrett. "Comparison of Spontaneous Mutation Rates of Normal and Chemically Transformed Human Skin Fibroblasts," *Cancer Res.*, 43:1650 (1983).
9. Cunningham, M.J., P. Kurian, and G. Milo. "Metabolism and Binding of Benzo[a]pyrene in Randomly-Proliferating Confluent and S-Phase Human Skin Fibroblasts," *Cell Biol. Toxicol.*, 5:155 (1989).

10. Kurian, P., S. Nesnow, and G.E. Milo. "Quantitative Evaluation of the Effects of Putative Human Carcinogens and Related Chemicals on Human Foreskin Fibroblasts," *Cell Biol. Toxicol.*, 6:171 (1990).

11. Barrett, J., T. Tsutsui, and P.O.P. T'so. "Neoplastic Transformation Induced by a Direct Perturbation of DNA," *Nature*, 294:229 (1978).

12. Barrett, J.C., and P.O.P. T'so. "Evidence for the Progressive Nature of Neoplastic Transformation in Vitro," *Proc. Natl. Acad. Sci.*, 75:3761 (1978).

13. Lehman, T.A., and C.C. Harris. "Oncogene and Tumor Suppressor Gene Involvement in Human Lung Carcinogenesis," in *Transformation of Human Epithelial Cells: Molecular and Oncogenetic Mechanisms*, G.E. Milo, B.C. Casto, and C.F. Shuler, Eds., (Boca Raton, FL: CRC Press, Inc., 1992), Chap. 10.

14. Milo, G., and J. DiPaolo. "Presensitization of Human Cells with Extrinsic Signals to Induced Chemical Carcinogenesis," *Int. J. of Cancer* 26:805–812 (1980).

15. Cunningham, M.J., The Induction and Inhibition of Benzo[a]pyrene Metabolism in Human Epidermal Keratinocytes and Dermal Fibroblasts, thesis submitted in partial fulfillment of Ph.D. requirement, 1985.

16. Lehman, T.A., Studies in Human Epithelial Cell Carcinogenesis, thesis submitted in partial fulfillment of Ph.D. requirement, 1987.

17. Kurian, P., A. Jeffrey, and G. Milo. "Preferential Binding of Benzo[a]pyrene Diol Epoxide to the Linker DNA of Human Foreskin Fibroblasts in S Phase in the Presence of Benzamide," *Proc. Natl. Acad. Sci.*, 82:2769 (1985).

18. Ribovich, M., P. Kurian, and G. Milo. "Specific BPDE-I Modification of Replicating and Parental DNA from Early S Phase Human Foreskin Fibroblasts," *Carcinogenesis*, 7:737 (1986).

19. Milo, G.E., and B.C. Casto. "Events of Tumor Progression Associated with Carcinogen Treatment of Epithelial and Fibroblast Compared with Mutagenic Events," in *Transformation of Human Epithelial Cells: Molecular and Oncogenetic Mechanisms*, G.E. Milo, B.C. Casto, and C.F. Shuler, Eds., (Boca Raton, FL: CRC Press, Inc., 1992), Chap. 11.

20. Milo, G.E., B.C. Casto, and S.L. Huang. "Relationship Between Bulky Carcinogen-DNA Modification: Carcinogenesis and Mutagenesis," in *Transformation of Human Diploid Fibroblasts: Molecular and Genetic Mechanisms*, G.E. Milo, and B.C. Casto, Eds., (Boca Raton, FL: CRC Press, Inc., 1990), Chap. 8.

21. Oldham, J.W., L.E. Allred, G.E. Milo, O. Kindig, and C.C. Capen. "The Toxicological Evaluation of the Mycotoxins T-2 and T-2 Tetraol Using Normal Human Fibroblasts in Vitro," *Toxicol. Appl. Pharmacol.*, 52:159 (1980).

22. Allred, J., J. Oldham, G.E. Milo, O. Kindig, and C. Capen. "Multiparametric Evaluation of the Toxic Responses of Normal Human Cells Treated in Vitro with Different Classes of Environmental Toxicants," *J. Toxicol. Environ. Health*, 10:143 (1982).

23. Lewis, J., and G. Milo. "Image Analysis and Mathematical Correlation Between Cell Number and Diameter of Colonies in Soft-Agar as a Measurement of Growth in Soft-Agar," *Teratogenesis Carcinog. Mutagen.*, 10:351 (1990).

24. Kun, E., E. Kirsten, G. Milo, P. Kurian, and H.L. Kumari. "Cell Cycle Dependent Intervention by Benzamide of Carcinogen Induced Neoplastic Transfor-

mation and in Vitro Poly(ADP-ribosyl)ation of Nuclear Protein in Human Fibroblasts," *Proc. Natl. Acad. Sci.*, 80:7219 (1983).

25. Milo, G., J. Oldham, R. Zimmerman, G. Hatch, and S. Weisbrode, "Characterization of Human Cells Transformed by Chemical and Physical Carcinogens, in Vitro," *In Vitro*, 17:719 (1981).

26. Zimmerman, R., and R. Little. "Characterization of Human Diploid Fibroblasts Transformed in Vitro by Chemical Carcinogens," *Cancer Res.*, 43:2181 (1983).

27. Baker, S.J., E.R. Fearon, J.M. Nigro, S.R. Hamilton, A.C. Preisinger, J.M. Jessup, P. van Tuinen, D.H. Ledbetter, D.F. Barkers, Y. Nakamura, R. White, and B. Vogelstein. "Chromosome 17 Deletions and p53 Gene Mutation in Colorectal Carcinomas, *Science*, 244:217 (1989).

28. Balaban, G., M. Herleyn, D. Guerry IV, R. Bartolo, H. Koprowski, W. Clark, and P.C. Nowell. "Cytogenetics of Human Malignant Melanoma and Premalignant Lesions, *Cancer Genetics and Cytogenetics*, 11:429 (1984).

29. Albino, A.P., A.N. Houghton, M. Ersinger, J.S. Lee, R.R.S. Kantor, A.H. Oliff, and L.J. Old. "Class II Histocompatibility Antigen Expression in Human Melanocytes Transformed by Harvey Murine Sarcoma Virus, (Ha-MSV) and Kersten MSV Retroviruses," *J. Expt. Medicine*, 164:1710 (1986).

PART II

Biological Effects of Low Level Exposure to Chemicals

CHAPTER 5

Nonmonotonic Dose-Response Relationships in Toxicological Studies

J. Michael Davis, Environmental Criteria and Assessment Office, Office of Health and Environmental Assessment, U.S. Environmental Protection Agency, Research Triangle Park, North Carolina

David J. Svendsgaard, Biostatistics Branch, Research Support Division, Health Effects Research Laboratory, U.S. Environmental Protection Agency, Research Triangle Park, North Carolina

INTRODUCTION

As Voltaire once said, "Before you speak to me you must define your terms." That precept seems quite appropriate in the case of nonmonotonic dose-response relationships. "Nonmonotonic" refers to a quantitative relationship between two variables that is characterized by a bidirectional change in a dependent variable as the independent variable is changed in only one direction. In the context of dose-response curves, a nonmonotonic relationship describes instances in which treatment by a toxic agent is accompanied by a response first in one direction, then the other, as illustrated in Figures 5.1a and 5.1b. Note that the term "dose-response" is used broadly; that is, the term "dose" includes exposure concentrations as well as directly administered or delivered doses, and the term "response" encompasses continuous effects as well as quantal or discrete responses.

Descriptive terms such as *U-shaped*, *J-shaped*, and *umbrella-shaped* are sometimes applied to these types of curves. These labels should not be taken too literally because, depending on the nature of the endpoint being measured, such dose-response curves could actually appear as upside-down *U*'s or *J*'s (Figure 5.1b). Other terms, including *biphasic*, *inverted*, and *paradoxical* dose-response, have also been invoked to describe this type of nonmonotonic relationship. In addition, terms such as *hormesis*, *sufficient challenge*, and the *Arndt-Schulz Law* are sometimes applied to nonmonotonic dose-response relationships that demonstrate a low-level stimulatory effect of a toxicant. One problem with the latter terms is that they are often

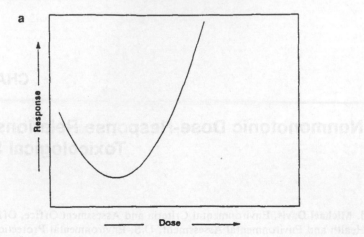

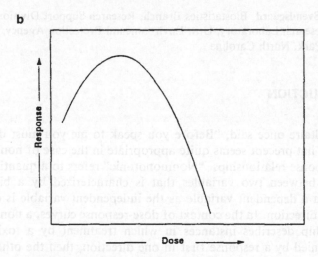

Figure 5.1a, b. Nonmonotonic dose-response curves.

applied in a manner that suggests they are more meaningful than they really are; that is, they are frequently used as if they provide some sort of explanatory concept, when in fact they are only denoting phenomena for which, in general, underlying mechanisms are not adequately understood. There should be no illusion that any of these terms has any fundamental explanatory value.

In a previous paper,[1] we discussed some aspects of terminology and attempted to distinguish certain types of U-shaped dose-response relationships based on an informal examination of the published literature in toxicology and environmental health. Our approach could be compared to that of a 17th-century naturalist who might have happened upon and collected biological specimens without necessarily having a well-developed classification scheme based on biological theory or systematics. Nevertheless, we found it possible, and thought it useful, to distinguish among certain types of nonmonotonic relationships already known to toxicologists.

Pharmacological agents are generally recognized as producing desirable effects at therapeutic dose ranges and undesirable toxic effects when the dose is too high. In an oft-quoted writing from the 16th century, Paracelsus stated that the dose determines whether a substance acts as a poison or as a therapeutic agent.[2] Unfortunately, for most chemicals in the environment, this aphorism provides little useful guidance in assessing potential health risks to a population, especially when there is no basis to assume that the chemicals in question necessarily have therapeutic effects or when the chemicals are known to have definite toxic properties (even if in conjunction with some desirable property).

Another well-known type of nonmonotonic dose-response relationship is illustrated by nutritionally essential substances. As shown in Figure 5.2, an essential element can improve physiological functioning within an optimal range of dietary intake and can produce toxicity at intake levels that are too high. The reference point in these cases is that of nutritional deficiency; that is, the peak in the curve describes not so much an enhancement of function as the absence of dysfunction associated with a dietary deficiency of the substance. Although the nonmonotonic relationships under consideration here share some features with nutritionally essential elements, it would be obviously fallacious to assume that any agent producing a nonmonotonic relationship is nutritionally essential. Indeed, relative to the very large number of natural and xenobiotic agents in existence, the known essential elements are quite few in number.

Yet another type of nonmonotonic curve familiar to toxicologists concerns the time course of a response to an agent (Figure 5.3), but of course this is not a dose-response curve. One might, however, classify such data as

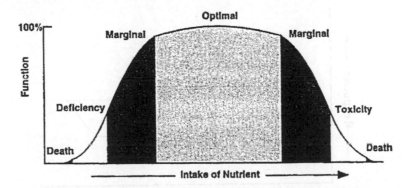

Figure 5.2. Nonmonotonic relationship between essential nutrient intake and physiological function. Source: Mertz.[13] Reproduced with permission of the author and the American Association for the Advancement of Science.

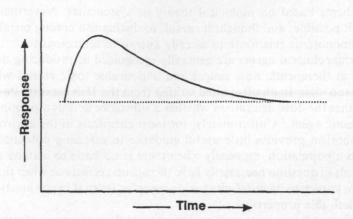

Figure 5.3. Toxicological response as a function of time.

paradoxical if there were an overshoot in the response after returning to baseline, as illustrated in Figure 5.4.

This chapter focuses on nonmonotonic toxicological dose-response relationships that suggest a bidirectional effect in a measured endpoint as the dose of a toxic agent is increased. Of particular interest are cases in which a low dose of an agent produces effects that appear to be in the direction of reduced dysfunction (Figure 5.5a) or enhancement of normal function (Figure 5.5b). It must be emphasized that one should be very cautious in interpreting such low-dose effects as necessarily beneficial, which is why we describe such effects as *apparent* improvements in function. We therefore prefer to suspend evaluative judgment and use the term *U-shaped* to describe this type of nonmonotonic dose-response relationship, primarily

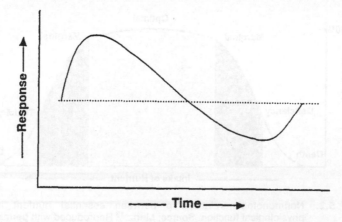

Figure 5.4. Toxicological response as a function of time, with overshoot beyond baseline.

a

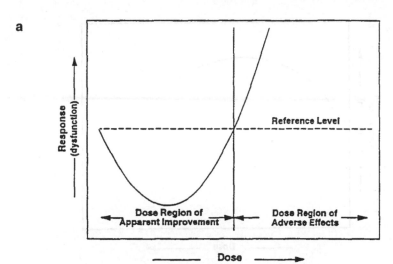

b

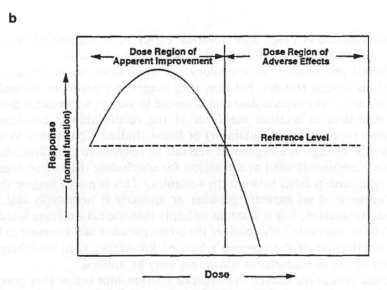

Figure 5.5a,b. U-shaped dose-response curves illustrating apparent (a) reduced dysfunction and (b) enhanced function.

because the term is obviously only a descriptive label and has no theoretical or value implications.

Another type of nonmonotonic curve, in which lower doses produce expectable toxic responses, whereas higher doses produce a return to control or normal function level (Figure 5.6), has been noted by others.[3-5] For the purposes of this discussion, the latter type of dose-response relationship

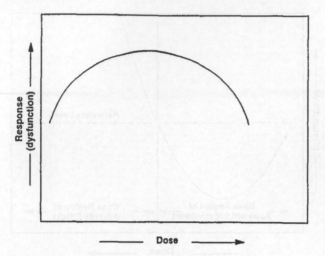

Figure 5.6.   High-dose, reversal-type, dose-response curve.

will be designated as a high-dose "reversal" effect, to distinguish it from the low-level paradoxical U-shaped relationship described above.

U-shaped relationships in toxicology and environmental health are of interest for several reasons. For one, they suggest a paradox or anomaly, because one usually expects dose and response to vary in a consistent direction, regardless of whether the form of the relationship is curvilinear (including supralinear and sublinear) or linear. Indeed, consistency in the direction of change in a dependent variable in relation to an independent variable is commonly cited as a criterion for concluding that a true cause-effect relationship exists between the variables.[6] This is not to suggest that every instance of an apparent paradox or anomaly is necessarily real or without explanation, nor is it meant to imply that the net outcome is truly beneficial or desirable.[1] We consider the terms paradox and anomaly to be more a reflection of the observer's lack of knowledge than an inherent property of the phenomena to which they may be applied.

Another reason for interest in U-shaped relationships is that they potentially pose complications for risk assessment methods that are based on the identification of a no-observed-adverse-effect level (NOAEL) or lowest-observed-adverse-effect level (LOAEL). For example, as illustrated in Figure 5.7, in the absence of data for doses 1 and 2, one would probably conclude that the NOAEL in these hypothetical data was at dose 3. With the data for doses 1 and 2, the task of selecting a NOAEL or LOAEL could be rather more difficult.

If U-shaped dose-response relationships occur only very rarely, they would be of little practical consequence to environmental toxicology or risk assessment. On the other hand, if the incidence of such relationships is more

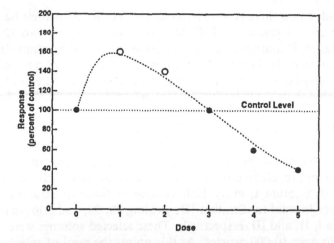

Figure 5.7. Hypothetical toxicity data illustrating how presence or absence of data for doses 1 and 2 may influence selection of LOAEL and NOAEL (see text).

than trivial, then the issues they raise presumably deserve greater consideration.

The objective of this study was to quantify the incidence of nonmonotonic relationships in toxicological studies, with a particular focus on U-shaped dose-response relationships as defined above.

METHODS

Our plan was to sample the published literature to determine the frequency of occurrence of results in experimental toxicology reports that suggested nonmonotonicity. Because the focus was toxicants, the U.S. Environmental Protection Agency's Integrated Risk Information System (IRIS)[7] database of reference doses (RfDs) was used to define a population of appropriate journals to be sampled. An RfD is defined as "an estimate (with uncertainty spanning perhaps an order of magnitude) of a daily exposure to the human population (including sensitive subgroups) that is likely to be without an appreciable risk of deleterious effects during a lifetime."[7] An RfD is typically derived by taking a NOAEL or a LOAEL from a critical study and dividing by uncertainty factors. Thus, the IRIS database provides a source from which to identify some of the most relevant journals from the standpoint of health risk assessment.

An examination of IRIS yielded 43 different journals that had been cited for 88 distinct, critical studies in deriving RfDs. (Books, private laboratory reports, and government documents were also the source for such studies, but were excluded from consideration here.) From this population of journals, a stratified statistical sampling of articles was undertaken. Journals were divided into three strata reflecting the frequency of their having been cited for critical studies in IRIS: Stratum I journals had 4 to 12 citations each; Stratum II journals had 2 to 3 citations each; and Stratum III journals had 1 citation each. Two neurotoxicology journals were also added to Stratum I, largely because neurotoxicology was under-represented in toxicology journals due to its comparatively recent emergence as a major subdiscipline of toxicology.

To enhance the efficiency and objectivity of sampling for articles relevant to toxicological risk assessment, optimal allocation sampling[8] was employed, based on the proportion of journal citations within each stratum. This approach resulted in the selection of every 7th volume of the journals in Stratum I, every 16th volume of those in Stratum II, and 1 volume per journal for Stratum III, yielding 33, 23, and 6 journal volumes for Strata I, II, and III, respectively. These selected volumes were estimated to contain over 10,000 articles. At this point, the pool of selected journal volumes was divided to create two groups. Only the first group was used for the present study; the second is available for replication purposes and may be the subject of a future report.

Several exclusion criteria were used to select the most appropriate types of journals and articles for consideration. A fundamental objective was to select articles that contained experimental toxicological results obtained with laboratory mammals. An article was not excluded if it also contained human data, but a selected article could not contain exclusively human data. All of the exclusion criteria are listed in Table 5.1. Some criteria (e.g., foreign language, published before 1950) resulted in the elimination of certain journals or journal volumes from further consideration.

Employing the criteria in Table 5.1, assistants examined the tables of contents of the selected journal volumes and, based on the titles of 1,836 articles contained in those volumes, classified the papers into three categories: 1,320 "no" articles were excluded for one or more of the reasons listed in Table 5.1; 431 "yes" articles did not appear to fit any of the exclusion criteria; and 85 "maybe" articles were difficult to judge from their titles alone. From these three sets, none of the "no" articles were considered

Table 5.1. Exclusion Criteria for Published Toxicology Articles

Lack of experimental laboratory animal data	Exclusive focus on: methodology
Lack of experimental laboratory animal data	Exclusive focus on:
animal data	methodology
Chemical/agent not identified in	diagnostics
article title	mechanisms
Abstract only	in vitro systems
Foreign language	cytotoxicity
Review article	genotoxicity
Letter to editor	mutagenesis
Studies of radiation, vibration, and	carcinogenesis
hypoxia	pharmacokinetics
Published before 1950	metabolism
	biomarkers

further; 100% of the "yes" articles were subjected to further examination; and 33% of the "maybe" articles were randomly selected for further study. At this stage, we had a pool of 459 photocopied articles. From preliminary statistical power calculations, it was decided to limit the initial examination to approximately the first 150 articles that were determined to meet the selection criteria. Assistants looked up each of the selected articles and read the abstract or scanned the article to decide if the report fit any of the exclusion criteria. If at this stage they judged the article usable, they obtained a photocopy of it and entered information about specified aspects of the report into a computer database. In this manner, 150 articles from 10 journals were incorporated into the study (Table 5.2).

Given the exploratory nature of this study, we wanted to avoid imposing criteria for nonmonotonicity that were too stringent. For example, for reasons that will be discussed further below, results did not have to be reported as "statistically significant at $p \leq 0.05$" to be considered U-shaped. On the other hand, a specifiable, objective criterion of acceptance was necessary. Therefore, for the first-cut selection process, a change of 5% or more in a response compared to the control level of response was considered sufficiently different to qualify initially for classification as a U-shaped relationship. Also, only one dose level needed to show a 5% "improvement" over control to meet this criterion, and this did not necessarily have to be the lowest dose group. A more stringent criterion requiring a difference of at least two standard errors of the mean was also applied subsequently, as well as in-depth evaluation of selected cases.

It was evident from the outset that careful reading of an article would be necessary to detect nonmonotonic dose-response relationships, because in many cases such data are not highlighted or even acknowledged by investigators. (A computerized key-word search would therefore have been futile for these purposes.) Indeed, nonmonotonic results are often "hidden" in tables or in the text of an article. For this reason, the search for nonmonotonic relationships was by no means limited to graphical plots of U-shaped curves; rather, the intent was to find information in any form that would

Table 5.2. Journals and Numbers of Articles Selected for Inclusion in Study

Journal	Number of Articles
Archives of Environmental Health	14
Agricultural and Food Chemistry	7
Environmental Health Perspectives	11
Fundamental and Applied Toxicology	22
Food Chemistry and Toxicology	9
Food and Cosmetic Toxicology	8
Journal of Environmental Pathology, Toxicology and Oncology	4
NeuroToxicology	15
Neurotoxicology and Teratology	13
Toxicology and Applied Pharmacology	47
TOTAL	150

reveal the nature of the relationship between an administered toxicant and the outcome(s) of such treatment. Thus, in tabulating and classifying dose-response relationships, every endpoint for which measurements were reported and could be related to one or more dose levels of an identified agent was considered a separate dose-response relationship. Consequently, a "small" study limited to examining effects in a single organ might contain several dose-response relationships. As a quality assurance provision, each article was examined by two or more persons: one or more assistants, at least one of the present authors, and in some cases two consultant toxicologists.

RESULTS

Of the 150 articles initially examined and collected by our assistants, we ultimately determined that 147 actually met all of the selection criteria for the study. (The three exclusions comprised a letter to the editor, a review article, and an experimental cancer study, each of which fit an a priori exclusion criterion, but was not readily apparent at first examination.) As shown in Table 5.3, 25 (17%) of the 147 articles contained some sort of indication of paradoxical effect or nonmonotonic dose-response relationship, 22 (15%) of which contained quantitative data, and three (2%) of which qualitatively described the results as paradoxical, although the form of the data as presented did not allow derivation of a dose-response curve. Another eight (5%) articles reported "reversal" type curves, where lower doses produced a toxic effect but at higher doses the response seemed to disappear. The data presented in another 84 (57%) of the 147 articles were either monotonic or could not be easily classified into any of the preceding categories (e.g., "W-shaped" functions where the response levels shifted back and forth across the control level as dose increased). In addition, 30 (20%) of the articles contained nonmonotonic curves that were time-related changes in the response to an agent but not dose-response relationships.

These classifications were not mutually exclusive. A given article might contain some data indicative of a monotonic relationship as well other data

Table 5.3. Number and Percentage of Data Pattern Classifications

Type	No. of Articles (n)	Percentage (n/147)
U-shaped dose-response	22	15%
Qualitatively paradoxical	3	2%
Subtotal	25	17%
High-dose reversals	8	5%
Other dose-responses	84	57%
Time-related responses	30	20%
TOTAL	147	99%[a]

[a]Total less than 100% because of rounding error.

indicative of a nonmonotonic relationship. In these cases, the classification of the article was hierarchical. That is, a U-shape took precedence over other types, and so on down the list of categories shown in Table 5.3.

This first-cut classification indicated that as many as approximately 17% of toxicology articles meeting our selection criteria contained data suggestive of paradoxical or U-shaped dose-response relationships. This is *not* to say that the incidence of such nonmonotonic relationships in toxicological studies is necessarily 17%, because a given article may contain multiple dose-response relationships for the same or different endpoints assessed in the study. Calculation of the incidence of U-shaped curves based on the totality of dose-response relationships was not feasible for the study, because it was estimated that it would have required examination of several thousand dose-response relationships in the 147 sampled articles. Instead, the incidence of U-shaped relationships was based on the total number of dose-response relationships in the 22 articles containing an indication of such nonmonotonic functions (Table 5.4). Of the 780 dose-response relationships reported in those 22 articles, 93 met our 5% rule for U-shapedness. By this criterion, 12% (i.e., 93/780) of the dose-response relationships were U-shaped in studies that had at least one U-shaped relationship.

Rarely were these U-shaped data reported as statistically significant. Only in two studies did the investigators' own statistical analyses indicate significant differences between a control response and the paradoxical response. These two studies reported a total of 54 dose-response relationships, of which two (3.7%) were significantly U-shaped. Of course, if the test applied by an investigator assumed monotonicity, as is most commonly the case, then a priori it would not have detected a statistically significant change in the "wrong" direction. It should also be noted that if a two-sided test of significance were employed, considerable loss of statistical power would result. Thus, one has the dilemma of choosing between a one-sided test to maintain statistical power and a two-sided test to detect effects in either direction.

Nevertheless, some degree of statistical assessment was necessary for our purposes. As a rough and easily applied "rule of thumb," we considered a difference of more than two standard errors of the mean to provide an indication of potential statistical significance. Because many articles did not report sufficient information to allow even this crude statistical analysis,

Table 5.4. Incidence of U-shaped Relationships in 22 Selected Studies

Criterion for U-shape	Number per Total[a]	Percentage
5% Difference	93/780	12
2 S.E.M.	22/180	12

[a]The denominator represents the total number of dose-response relationships for which the specified criterion could be applied, based on the published report of the data.

the incidence of U-shaped relationships was calculated based on 180 dose-response relationships in the only articles providing adequate statistical information. By this criterion, 22 (12%) of 180 dose-response relationships could be considered U-shaped (Table 5.4).

In addition to statistical significance, we considered the toxicological significance of the U-shaped findings. Twelve articles from the set of 22 that appeared to demonstrate a U-shaped relationship were selected and provided to a consulting firm that specializes in toxicological risk assessment. The firm was asked to provide a determination of the NOAELs and LOAELs for each study, with specific reference to those effects suggestive of a U-shaped curve, and to comment on the toxicological significance of the effects. A quote from the cover letter of the toxicologist who evaluated the articles summarizes his view of them: "Subtle changes in the shape of the dose-response curve were occasionally found; however, because many of these changes were slight and statistically nonsignificant (possibly reflecting biological and analytical variability), the toxicity assessment was based on the total spectrum of the dose-response function and on other endpoints examined in the study." Although it is certainly appropriate to consider all of the data available in conducting a risk assessment, it is nonetheless interesting that, despite explicit instructions to focus on "those dose-response functions that displayed 'atypical' shapes," especially U-shapes, the assessor tended to discount or otherwise ignore the anomalous effects.

In cases in which the effect in question was the primary outcome of the study, as in a report by Wood and Colotla[9] (see Figure 5.8), the assessor's evaluation still seemed a bit hesitant: "The toxicological significance of the increase in motor activity observed at dose levels of 560–1780 ppm is not

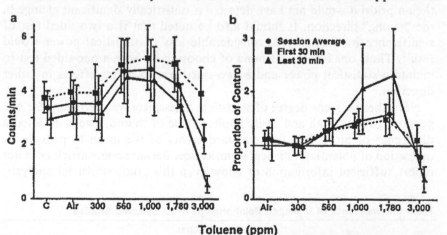

Figure 5.8a,b.   Effects of toluene on activity in mice expressed as (a) group mean (± 2 standard errors) locomotor activity counts and (b) mean proportion of control, with individuals serving as their own controls. Source: Wood and Colotla.[9] Reproduced with permission of the authors and Academic Press, Inc.

known. Currently, the criteria for defining an 'adverse' effect for neurotox-
icity, for dose-response assessment purposes, remain unclear." Nevertheless,
citing "a tendency among neurotoxicologists and behavioral toxicologists to
conclude that any behavioral effect, independent of the direction of the
effect produced by a nontherapeutic chemical constitutes a behaviorally
toxic effect," the writer concluded that 560 ppm was the LOAEL and 300
ppm was the NOAEL.

An independent academic toxicologist was also asked to review the same
articles and provide comments on the results of the studies. This reviewer
concluded that 5 of the 12 articles provided rather convincing evidence of
U-shaped effects, whereas 7 were not sufficiently persuasive, based on the
reported data. If the incidence of U-shaped relationships is calculated based
on the total number of dose-response relationships in the 5 positive articles,
41 (24%) of 171 relationships were U-shaped.

The cases that were rejected are also informative. For example, in a
behavioral toxicology study by Knisely et al.,[10] carbon monoxide (CO) was
administered intraperitoneally with various drugs to investigate their interactions as reflected by response performance on an operant conditioning
schedule. In graphs of the data for the various drugs, the effect of CO plus
the saline vehicle for the drug was also plotted (see hatched bars in Figure
5.9). In six separate experiments, CO and saline were administered as a
control condition. (Note that response rates shown as a "percentage of
control" refer not to the zero CO dose condition, but to the mean of three
previous days of training in the absence of CO, saline, or drugs.) In all six
of these replicate experiments, one or more of the CO doses produced
higher response rates than the zero-CO control condition. Apparently, neither of the two independent toxicologists considered these data to be sufficiently noteworthy or consistent to support a conclusion that the effect of
CO and saline on performance was U-shaped. We, on the other hand, were
struck by the degree of similarity across replications. Moreover, similar
examples of such excitatory effects at relatively low levels of exposure to
other neurobehavioral toxicants have been previously noted.[1] From a statistical standpoint, however, what can one conclude?

A post hoc test of the differences between the various treatment means
and the control mean in the "best case" of these replications (Figure 5.9d)
indicated that only the decline in response rate at the highest dose level was
significantly different from the control rate by Dunnett's test.[11] That is, the
elevated response rates were not found to be statistically different from the
control response rate. However, a modeling approach using nonlinear
regression provided another perspective on the data. A quadratic function
was fitted to the mean response rate versus the natural log of the nonzero
dose levels. (The log of dose was used because, among other reasons, it
expands the lower region of the dose scale, which is the area of primary
interest here; however, because there is no zero on a log scale, the model
applies only to nonzero doses and can be only roughly extrapolated to the

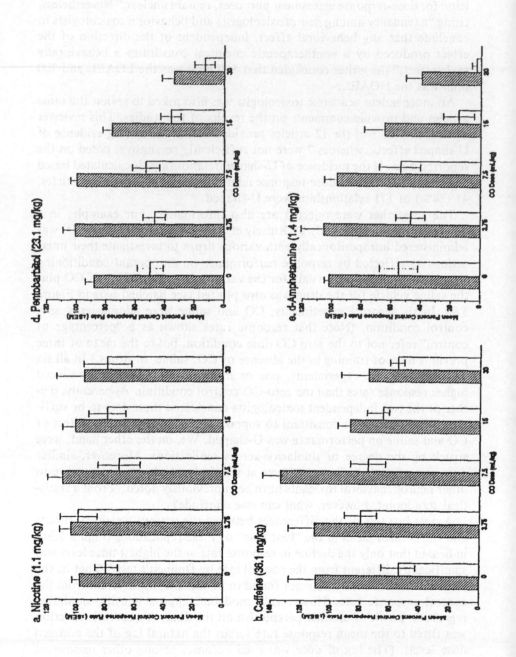

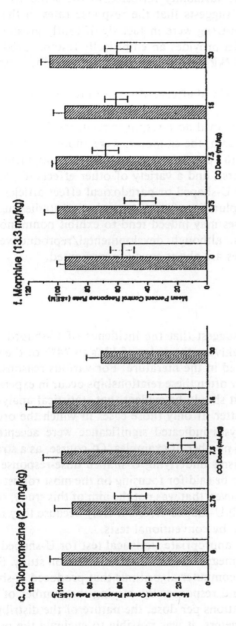

Figure 5.9, a–f. The effects of carbon monoxide (CO) with either saline (hatched bars) or various drugs (open bars). Source: Knisely et al.[10] Reproduced with permission of the authors and Pergamon Press, Ltd. For chapter 16

control level.) As shown in Figure 5.10, the regression model, which has been modified to include the variability reflected in the standard errors reported by Knisely et al.,[10] suggests that the response rates in the dose region from about 4 to 6.5 mL/kg were in fact significantly greater than controls. Note that this model provides an empirically testable prediction that a dose of about 5.5 mL/kg would produce an even higher response rate.

As noted above, neurobehavioral effects often seem to exhibit U-shaped relationships. This raised the question of whether certain types of endpoints more frequently manifest U-shaped dose-response relationships. Of the 22 articles containing U-shaped data, the endpoints were about evenly divided among neurotoxicological (6 papers), reproductive/developmental (5 papers), body weight (5 papers), and a variety of other effects (6 papers). All three of the qualitatively U-shaped or paradoxical effect articles dealt with behavioral effects. This plurality of neurotoxicological studies suggests that neurobehavioral processes may indeed tend to exhibit nonmonotonic dose-response characteristics, although developmental/reproductive and body weight/growth processes are about equally represented.

DISCUSSION

The results of this study suggest that the incidence of U-shaped dose-response relationships is roughly on the order of 12% to 24% of the dose-response relationships reported in the literature. For various reasons, it is difficult to state precisely how often these relationships occur in experimental toxicology studies. First, it should be evident that statistical analysis of such data is not a simple matter. If only those cases in which the original investigators' statistical analyses indicated significance were accepted as valid, there would be a trivial number of examples. Of course, as a strategy for investigating the mechanisms underlying U-shaped dose-response phenomena, there is something to be said for focusing on the most robust cases and disregarding others. However, that was not the aim of this study, so it is important not to dismiss small U-shaped effects merely because they fail to achieve statistical significance by conventional tests.

The problem of finding an appropriate statistical test for U-shaped data prompted further inquiry by means of a computer simulation study. Simulated data were generated by computer algorithms that produced U-shaped relationships between dose and response. By varying the number of dose levels, the number of observations per dose, the nature of the distribution of the data, and other parameters, it was possible to evaluate the performance of various statistical techniques in detecting a "true," albeit simulated, nonmonotonic relationship. This work may be the subject of a separate, more detailed report, but in brief it was evident that increasing the number of dose levels, increasing the number of observations per dose,

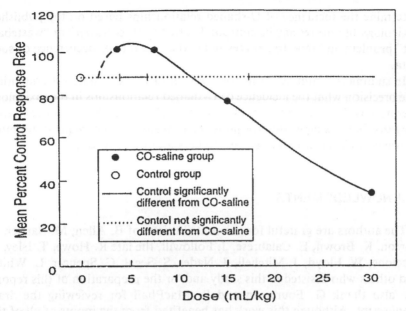

Figure 5.10. Model fitted to data from Knisely et al.[10] Quadratic function used to fit solid line to data, with dashed line representing the extrapolation toward zero dose.

decreasing the variability in the observations, and having normally distributed data improved the power of most tests to detect a U-shaped relationship.

The best overall performance seemed to be achieved with six or more dose levels and 30 "subjects" per group when analyzed with the permutation test.[12] However, further work is needed to determine how dose-spacing affects these results; also, other forms of statistical tests remain to be evaluated. In any event, it is clear that very few studies in the toxicology literature provide an adequate statistical basis for detecting a U-shaped dose-response relationship.

It is also evident that the phenomenological problem of simply recognizing U-shaped data makes it difficult to determine the incidence of these relationships. Human judgment has to be exercised in deciding whether or not data actually conform to a U-shaped pattern. Such judgment may be subject to subjective biases or other influences that tend to make a U-shaped relationship more, or less, evident to the eye. Even when reviewers accept the validity of U-shaped relationships, it may be difficult to achieve 100% agreement between different reviewers in the absence of clear-cut statistical criteria.

Another important consideration in interpreting the present findings is that there very likely is a bias against publishing, or even submitting for publication, "anomalous" data of the type in question. If so, the attempt to determine the incidence of U-shaped relationships based on the published toxicology literature may be confounded by the "file-drawer" or "wastebasket" problem and tend to *under*estimate the true rate of occurrence of such data.

In summary, a number of factors make it difficult to state with quantitative precision what the incidence of U-shaped relationships in the toxicology literature is. Nevertheless, our estimate of a 12% to 24% rate of occurrence suggests that U-shaped dose-response relationships occur frequently enough to warrant closer attention by researchers and risk assessors.

ACKNOWLEDGMENTS

The authors are grateful for the contributions of B. Allen, M. Barrier, J. Barton, K. Brown, E. Calabrese, J. Followill, the late R. Howe, T. Isley, J. Liccione, W. Lloyd, J. Mitchell, J. Nader, S. Segal, C. Spencer, L. White, and others who assisted in this study and/or the preparation of this report. We also thank G. Foureman and R. MacPhail for reviewing the draft manuscript. Although this work has benefited from the inputs of all of the above persons, none of them bears any responsibility for the views expressed here, nor do the views contained in this chapter, which are solely those of the authors, necessarily reflect the policies or views of the U.S. Environmental Protection Agency.

REFERENCES

1. Davis, J.M., and D.J. Svendsgaard. "U-shaped Dose-Response Curves: Their Occurrence and Implications for Risk Assessment," *J. Tox. Environ. Health*, 30:71 (1990).
2. Deichmann, W.B., D. Henschler, B. Holmstedt, and G. Keil. "What is There That Is Not Poison? A Study of the Third Defense by Paracelsus," *Arch. Toxicol.*, 58:207 (1986).
3. Nelson, B.K., Letter to the Editor: "Dose/Effect Relationships in Developmental Neurotoxicology," *Neurobehav. Toxicol. Teratol.*, 3:255 (1981).
4. Tachibana, T., Commentary: "Instability of Dose-Response Results in Small Sample Studies in Behavioral Teratology," *Neurobehav. Toxicol. Teratol.*, 4:177 (1982).
5. Piegorsch, W.W., F.K. Zimmermann, S. Fogel, S.G. Whittaker, and M.A. Resnick. "Quantitative Approaches for Assessing Chromosome Loss in Saccharomyces cerevisiae: General Methods for Analyzing Downturns in Dose Response," *Mutat. Res.*, 224:11 (1989).
6. Davis, J.A. *The Logic of Causal Order*. Sage University Paper series on Quantitative Applications in the Social Sciences, Series No. 07–055. Sage Publications, Beverly Hills, CA, 1985.
7. IRIS, Integrated Risk Information System [on-line data base], Cincinnati, OH:

U.S. Environmental Protection Agency, Office of Health and Environmental Assessment, Environmental Criteria and Assessment Office, 1990.

8. Cochran, W.G., *Sampling Techniques*, 2nd edition, (New York: John Wiley and Sons, Inc., 1963), Sect. 5.10.

9. Wood, R., and V. Colotla. "Biphasic Changes in Mouse Motor Activity During Exposure to Toluene," *Fund. Appl. Toxicol.*, 14:6 (1990).

10. Knisely, J.S., D.C. Rees, and R.L. Balster. "Effects of Carbon Monoxide in Combination with Behaviorally Active Drugs on Fixed-Ratio Performance in the Mouse," *Neurotox. Teratol.*, 11:447 (1989).

11. Dunnett, C.W. "A Multiple Comparison Procedure for Comparing Several Treatments with a Control," *J. Am. Stat. Assoc.*, 50:1096–1121 (1955).

12. Pitman, E.J.G. "Significance Tests Which May Be Applied to Samples from Any Populations," *J. Royal Statist. Soc. B.*, 4:119 (1937).

13. Mertz, W., "The Essential Trace Elements," *Science*, 213:1332 (1981).

CHAPTER 6

Biostatistical Approaches to Low Level Exposures

David W. Gaylor, National Center for Toxicological Research,
U.S. Food and Drug Administration, Jefferson, Arkansas

INTRODUCTION

Hormesis will be examined from the standpoint of health risk assessment in this chapter. Conditions will be discussed where low-dose exposures result in reduced disease risk. This implies the existence of a zero equivalent dose (ZED) where the incidence of disease is equal to the ambient or spontaneous background incidence. In this case, the extrapolation of risk estimates from higher doses should be directed to the ZED, more specifically to a lower confidence limit of the ZED, rather than to zero dose. This will result in lower estimates of risk at low doses.

The primary step in risk estimation is the development of the relationship between the incidence of disease and dose. Dose is used in a very general sense, but refers to the concentration of the active toxicant at the target tissue site. The active toxicant may be the administered chemical or a metabolite of the chemical. At low doses where the kinetics of chemical absorption, metabolism, and distribution are likely to be linear, the target tissue dose may be proportional to the administered dose. In such cases, risk estimates may be based upon the administered dose.

Davis and Svendsgaard[1] discuss problems associated with identifying and estimating hormetic effects. The presence of hormesis implies the existence of a U-shaped dose-response (incidence) curve, Figure 6.1. Note that dose is not necessarily the administered dose; rather, dose is the concentration of the active chemical at the target tissue site. The incidence is the probability (proportion of individuals) with a disease or adverse effect. The ambient (spontaneous background) dose is the dose already present due to other environmental exposures which account for the background incidence of disease. The ZED is the dose that produces a disease incidence equal to the ambient or spontaneous background incidence. Note that the ambient

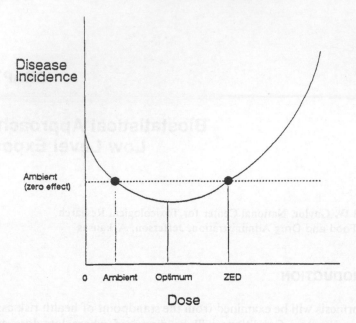

Figure 6.1. U-shaped hormetic relationship between disease incidence and dose.

(background) dose is often greater than zero due to endogenous and/or other exogenous factors. This would be the case for essential nutrients where both low levels and high levels can lead to adverse effects. For a chemical foreign to the body, the effective ambient dose equals zero if there is no other environmental exposure. The optimum dose is the dose where the incidence of disease is a minimum. A decrease in dose below the optimum dose results in an increase in the incidence of disease. Note that *a hormetic effect can only be obtained if the ambient (background) dose is less than the optimum dose. If the ambient dose is equal to or greater than the optimum dose, any additional increment of dose will result in an increased incidence; i.e., there is no threshold dose even though there is an underlying hormetic process, Figure 6.2.*

Another condition is likely to be common that may lead to misleading results. The ambient dose in a laboratory setting may be different than the ambient dose in the human environment. The ambient laboratory dose is the effective dose of the active chemical at the target tissue site in control animals without the addition of an administered dose of the chemical under investigation. The ambient environmental dose is the effective dose present of the active chemical at the target tissue site in humans without additional exposure to the chemical under investigation. *If in an uncontrolled natural environment the ambient environmental dose is greater than the ambient laboratory dose, then the addition of a dose which is beneficial in the laboratory may result in an increased risk in the environment, Figure 6.3.*

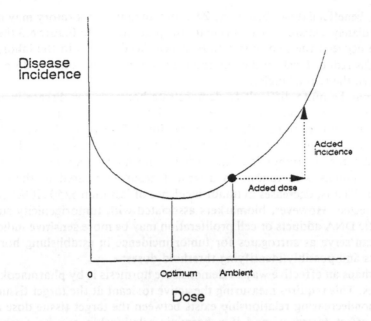

Figure 6.2. No threshold dose when the ambient dose exceeds the optimum dose.

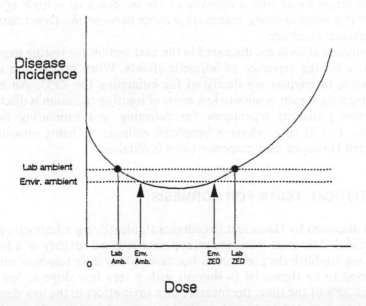

Figure 6.3. Added beneficial environmental dose less than added beneficial laboratory dose.

Thus, beneficial doses below the ZED obtained in a laboratory may not be immediately transferable to an environmental setting. In Figure 6.3 the dose range between the two dots results in beneficial effects in the laboratory, whereas reduced risk in the environment occurs over a smaller dose range between the two triangles.

It may be quite difficult to demonstrate hormesis for disease incidence because of the imprecision of the proportion of diseased animals in typical bioassays. For example, the current National Toxicology Program bioassays utilize 50 animals per dose group. To demonstrate a statistically significant decrease in tumor incidence (due to hormesis) requires at least five fewer animals with tumors in a treated group compared to the control group. That is, decreases in tumor incidence of less than 5/50 (10%) cannot be detected. However, biomarkers associated with tumorigenicity such as specific DNA adducts or cell proliferation may be more sensitive indicators and can serve as surrogates for tumor incidence in establishing hormetic effects and possibly identifying threshold doses.

Perhaps an effective way to demonstrate hormesis is by pharmacokinetic studies. This requires measuring the active toxicant at the target tissue site. If a nondecreasing relationship exists between the target tissue dose and a toxic effect (disease), and if a hormetic relationship can be established between the target tissue dose of the active toxicant (parent chemical or metabolite) and the administered dose, then doses below the ZED of the administered dose will result in a decrease of the incidence of a toxic effect. That is, *at low administered doses that result in a decrease in the active dose at the target tissue site, a decrease in the incidence of a toxic effect will occur if a nondecreasing relationship exists between the target tissue dose and disease incidence.*

Statistical criteria are discussed in the next section for testing experimental data for the existence of hormetic effects. When hormesis is present, statistical techniques are described for estimating the ZED. An example relating weight gain in mice to low levels of ionizing radiation is discussed to illustrate statistical techniques for detecting and estimating hormetic effects. In this case, where a beneficial endpoint is being measured, an inverted U-shaped dose-response curve is obtained.

STATISTICAL TESTS FOR HORMESIS

As discussed by Davis and Svendsgaard[1], identifying a hormetic effect is not trivial. The mere existence of apparent reduced toxicity at a low dose does not establish the presence of hormesis. Many dose response curves are observed to be sigmoidal (S-shaped) with a very low slope at low doses. About 50% of the time, the measure of a toxic effect in the low dose range would be expected to be below the background toxicity observed in the controls. The question becomes, how large a decrease in toxicity at low doses is required to obtain adequate experimental (statistical) evidence to support the existence of hormesis.

A statistically significant reduction in toxicity at a low dose compared to a control group is not a sufficient demonstration of hormesis, and a possible explanation is that the toxicity observed in the control group is abnormally high. Perhaps the possibility of abnormally high controls might be reduced by using historical control values. However, the variability of results in bioassays for developmental effects shown by Holson et al.[2] and for tumorigenicity, as discussed by Gaylor et al.[3], limit the use of historical control values and generally the concurrent control results must be used.

To reduce the problem of abnormal toxicity in the control group, the control toxicity can be estimated from dose response data. Initially, one could fit a smooth monotonic dose response curve to the bioassay data. If the observed toxicity at a low dose is significantly less than the toxicity estimated by the fitted nonhormetic dose response curve, then there is statistical evidence to reject the null hypothesis of no hormesis. Such a test should be conducted as a two-sided test. Also, a high level of statistical significance, e.g., $P < 0.01$ should probably be required to avoid concluding that hormesis exists when it doesn't exist.

A good statistical fit of a hormetic dose response curve to bioassay data does not necessarily constitute proof of hormesis. This will be demonstrated later in this chapter. Because of the variability of quantal bioassay data for the incidence of disease, several different types of dose response curves can be used to fit experimental bioassay data equally well. Statistical goodness-of-fit tests can be used to reject dose response curves, but bioassay data generally are not good enough to discriminate between several plausible dose response curves.

Perhaps the best evidence for hormesis from a single set of dose response data is to show a statistically significant reduction in the error sum of squares when a hormetic curve is used. The first step is to find a best fitting smooth monotonic (nonhormetic) curve from among several biologically plausible dose response curves. If a hormetic dose response curve can be fit to the data which results in a statistically significant improvement in the fit, as measured by the reduction in the error sum of squares, then there is evidence for hormesis. Again, the statistical significance level should probably achieve a high level, e.g., $P < 0.01$ in order to avoid false acceptance of hormesis. The statistical test can be performed as follows. Suppose there are n data points providing measures of toxicity as a function of dose. Suppose the best fitting nonhormetic curve examined requires fitting p constants with a sum of squares of error of SSE. The calculation of the sum of squares for error is given by many statistical computing packages and described in standard statistical textbooks, e.g., Snedecor and Cochran.[4] Suppose a hormetic dose response curve requires fitting $q > p$ constants with a sum of squares of error of SSE_H. A test for the reduction in the error sum of squares due to fitting a hormetic curve is given by the F ratio test

$$F = \frac{(SSE - SSE_H) / (q - p)}{SSE_H / (n - q)} \qquad (1)$$

with $(q - p)$ and $(n - q)$ degrees of freedom. This test requires that the nonhormetic model is a subset (special case) of the hormetic model. A large value of F indicates an improved fit with the hormetic dose response curve. The size of F required for statistical significance can be obtained from F tables given in most standard statistical textbooks or statistical tables, e.g., Beyer[5].

Because of the variability in experimental data, it may be difficult to detect hormesis when it is present. Perhaps the best indication of hormesis is demonstrated by the reproducibility of results by different investigators at different laboratories. Even though no single set of data may provide convincing evidence of hormesis, consistency of effects in two or more studies may provide adequate evidence for hormesis.

The carcinogenic potency database compiled by Gold et al.[6] provides bioassay dose response data for approximately 3,000 chronic experiments of 770 chemicals. There were no convincing cases of hormesis noted in this large database. There may be several reasons for this. As stated before, the variability in typical tumor bioassay data generally makes it impossible to detect reductions in tumor incidence of less than 10%. A number of the reported bioassays only contained a single dose and controls, which were not considered here for hormesis. Most bioassays test at doses above one-tenth of the maximum tolerated dose, which hardly can be considered low doses. Obviously, reductions could only be noted in tumor sites that have large spontaneous background incidence in control animals, e.g., liver tumors or lymphoma in some strains of male mice. Hormesis for rare tumors would not be noticed. Bioassays which demonstrated downward trends in tumor incidence with increasing doses, but did not exhibit a U-shape dose response were not considered to be hormesis because they could be the result of: (1) chemotherapeutic effects, (2) competing risks at higher doses killing the animal before a tumor at a particular site had time to develop, or (3) reduced caloric intake at higher doses resulting in lowered tumor incidence. However, it is somewhat surprising that in such a large database there did not appear to be hormetic effects, even if only due to random variation in experimental results.

Bioassay data reported in *Teratology*, Vol. 1 (1968) through Vol. 40 (1990) were examined for hormetic effects of embryo/fetal toxicity. Data from 120 experiments on 93 chemicals showed no convincing demonstration of hormesis. Again, some of the reasons for failure to observe hormesis are the same as noted previously for tumorigenicity studies. In addition, the spontaneous background incidence of malformations appears to be near zero in most strains used in bioassays, making detecting of reduction in the incidence of major birth defects nearly impossible. Even if a reduction in the incidence of malformations were observed it might be the result of higher death rates of malformed fetuses. The background incidence of dead and/or resorbed fetuses in some strains might be high enough to offer an opportunity to detect large reductions due to hormesis.

ESTIMATION OF THE ZERO EQUIVALENT DOSE (ZED)

Downs[7] provides a discussion of conditions and dose response curves that might describe hormesis. For purposes of discussion here, only the linear-quadratic dose response curve is considered

$$y = b_0 + b_1 d + b_2 d^2 \qquad (2)$$

where y is the measure of toxicity, d is dose, and b_0, b_1, and b_2 are constants to be estimated from the data. At least for incidence data, the linear-quadratic model often provides an adequate fit, and certainly a first approximation to more sophisticated dose response curves.

When the sign of b_1 is negative and the sign of b_2 is positive, Eq. 2 is U-shaped with a minimum (optimum) at a dose $d = -b_1/2b_2$. In such cases, Eq. 2 might represent two different mechanistic processes. One, a detoxification process or homeostatic upregulation of a cellular defense mechanism (e.g., antioxidation, DNA repair, and stress proteins) for which toxicity decreases with a negative slope, b_1, which dominates at small doses. The second mechanistic process may be a toxic process described by $b_2 d^2$ which dominates at high doses. The zero equivalent dose (ZED) occurs when toxicity equals the toxicity at the background dose. If the ambient background dose is set equal to zero, dose represents added dose. The ambient background toxicity is $y = b_0$ and the ZED occurs where $y = b_0 = b_0 + b_1 d + b_2 d^2$ or

$$ZED = -b_1 / b_2. \qquad (3)$$

This represents a threshold dose below which toxicity is reduced, which can be estimated from the data. Snedecor and Cochran[4] provide least squares estimates of b_1 and b_2 from which the ZED can be estimated. Because of experimental variation and inadequacy of Eq. 2 to accurately describe the true dose response curve, there is uncertainty in the estimates of b_1 and b_2 as measured by their variances, V_1 and V_2, respectively.

The variance of ZED is approximately

$$V_Z \sim (V_1 + Z^2 V_2 + 2ZV_{12}) / b_2^2 \qquad (4)$$

where $Z = ZED$ and V_{12} is the covariance of the estimates of b_1 and b_2. Then, an approximate lower confidence limit on the ZED is given by

$$L_Z = ZED - t\sqrt{V_Z} \qquad (5)$$

where t is a standard t-value with $(n - 3)$ degrees of freedom corresponding to the desired level of confidence. Hence, L_Z could be used as a threshold dose for risk assessment, below which no increase in toxicity above ambient background toxicity is likely under the experimental laboratory conditions.

In the previous discussion, y represented a measure of toxicity, e.g., disease incidence, where the sign of b_1 is negative and the sign of b_2 is positive, giving a U-shaped dose response curve. Hormesis can also be represented where y is the measure of a beneficial effect. In this case, an inverted U-shaped curve is obtained with low doses and high doses resulting in less beneficial effects which may eventually achieve adverse levels. For this case, the sign of b_1 in Eq. 1 is positive and the sign of b_2 is negative. This may represent a beneficial effect with a positive slope of b_1 which dominates at low doses and an independent second mechanism which reduces the beneficial effect by $b_2 d^2$ that dominates at high doses. The maximum (optimum) beneficial effect occurs at a dose $d = -b_1/2b_2$. The ZED is estimated by Eq. 3 with the approximate variance given by Eq. 4. The lower confidence limit on the estimate of ZED provides a threshold dose for risk assessment, below which the beneficial effect is not likely to be less than the ambient background level under the experimental laboratory conditions. An example of this case is given in the next section.

EXAMPLE

Data collected by Morris et al.,[8] which are presented in Luckey,[9] show the average weight gain of mice exposed to ionizing radiation relative to unexposed control mice of 1.00, 1.03, 1.05, 1.13, 1.05, 1.12, and 0.86 at doses of 0, 0.125, 0.25, 0.5, 1, 2, and 8 rads per day, respectively. It is assumed that the ambient background dose is zero, or at least is negligible, so that the administered dose is used. A least squares fit of a monotonic quadratic with no threshold provides a good fit to the data

$$W = 1.065 - 0.003143R^2 \qquad (6)$$

where W is the average relative weight gain and R is rads per day. This equation is represented by the dashed line in Figure 6.4. Note that this model estimates the maximum relative weight gain of 1.065 occurring with no radiation (assuming background levels of ionizing radiation to be negligible in the laboratory). A linear-quadratic model, Eq. 2, allowing for hormesis gives least squares estimates of

$$W = 1.029 + 0.0679R - 0.011137R^2. \qquad (7)$$

This curve is shown as the solid curve in Figure 6.4.

The next step is to determine if the data justify using the hormetic curve

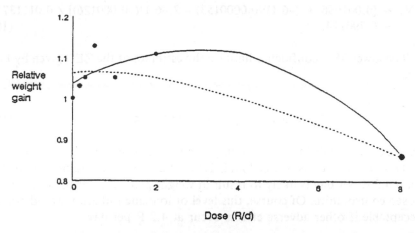

Figure 6.4. Average weight gain of mice exposed to ionizing radiation relative to unexposed controls (data from Morris et al.[8] and reported by Luckey[9]).

given by Eq. 7 rather than the nonthreshold monotonic curve given by Eq. 6. The sum of squares for error for the quadratic curve given by Eq. 6 is SSE = 0.01471 and the linear-quadratic hormetic curve in Eq. 7 the error sum of squares is SSE_{11} = 0.00707. The F-test for the reduction in the error sum of squares due to fitting the hormetic curve is from Eq.1

$$F = \frac{(0.01471 - 0.00707) / (3 - 2)}{0.00707 / (7 - 3)} = 4.32 \tag{8}$$

where F has 1 degree of freedom for the numerator and 4 degrees of freedom for the denominator. This F ratio is significant at the $P < 0.11$ level. There is an 11% chance that an F ratio of this size could have arisen by chance alone. Hence, there is not convincing evidence from these data that low levels of ionizing radiation increase weight gain in mice. If this were the only available data, there would be no justification for accepting hormesis. However, other investigators, e.g., Lorenz et al.,[10] have shown a similar effect. Hence, there is additional evidence to consider hormesis and to estimate the ZED.

The estimate of ZED from Eq. 3 is

$$ZED = -(0.0679) / (-0.011137) = 6.1 \text{ R/d} \tag{9}$$

where the relative weight gain is W = 1.029 and is equal to the estimate of the weight gain for the controls. The variances obtained from the least squares fit are V_1 = 0.001066, V_2 = 0.0000153, and V_{12} = -0.000126. The variance of the ZED is approximately

$$V_z \sim [0.001066 + (-6.1)^2(0.0000153) - 2(-6.1)(-0.000126)] / 0.011137^2$$
$$\sim 0.790114. \tag{10}$$

The lower 95% confidence limit on the estimate of the ZED given by Eq. 5 is

$$L_z = 6.1 - 2.13 \sqrt{0.790114} = 4.2 \text{ R/d}, \tag{11}$$

where the one-tailed t corresponding to a 95% confidence level with $(n - q)$ = $(7 - 3)$ = 4 degrees of freedom is 2.13. Thus, these data indicate that less than 4.2 R per day is likely to result in weight gains greater than the unexposed control mice. Of course, this level of ionizing radiation would not be acceptable if other adverse effects occur at 4.2 R per day.

SUMMARY

Hormesis is examined from the standpoint of risk assessment. Conditions are presented where hormesis results in reduced disease risk at low doses. This implies the existence of a zero equivalent dose (ZED) where the incidence of disease is equal to the ambient background incidence. In this case, the extrapolation of risk estimates from higher doses should be directed to the ZED, more specifically to a lower confidence limit estimate of the ZED, rather than to zero dose. This will result in lower estimates of risk at low doses.

Hormesis implies a U-shaped relationship between the incidence of a disease (or an adverse effect) and dose of the active toxicant at the target tissue site. The ambient background dose may be greater than zero due to endogenous and/or exogenous factors. A hormetic effect can only be obtained if the ambient (background) dose is less than the optimum dose. A beneficial dose less than the ZED obtained in a laboratory may not be beneficial in an environment that has a higher ambient dose. If the ambient dose is greater than the optimum dose, the addition of any dose, no matter how small, will result in an increased incidence of disease. Under this condition there is no threshold dose even though there may be an underlying hormetic process.

It is unlikely that hormesis can be detected in typical bioassays that identify only the proportion of diseased animals because of the experimental variability of this type of observation. Hormesis is more likely to require more precise measures of biomarkers which are associated with disease, or pharmacokinetic studies that demonstrate reductions in the active toxicant at the target tissue site for low doses of administered chemicals.

A statistically significant reduction in toxicity at a low dose compared to a control group is not a sufficient demonstration of hormesis, because an equally likely explanation may be an abnormally high incidence of disease in the control group. Historical control data are not likely to be useful. A model-based estimate of the control incidence may be used for comparison with low dose effects.

A good statistical fit of a hormetic dose response curve to bioassay data does not necessarily constitute proof of hormesis. However, statistical tests can be employed which show if a hormetic curve provides a statistically significant improvement in fit compared to a nonthreshold monotonic curve. A test was proposed here which is based on the F ratio. Because of experimental variability, hormesis may be difficult to demonstrate in a single study. Perhaps the best indicator of hormesis is provided by the reproducibility of results by different investigators at different laboratories.

When convincing evidence for hormesis exists, a U-shaped curve can be fit to bioassay data. Then, the ZED can be estimated where the disease incidence is equal to the ambient background incidence. Further, a lower confidence limit can be established for the estimate of the ZED which could be used as a threshold dose for risk assessment. It is likely that there will be no increase in toxicity above the ambient background incidence at doses below this lower limit, under the experimental laboratory conditions. Similarly, a beneficial range of doses can be estimated when a beneficial effect displays an inverted U-shaped dose response curve.

REFERENCES

1. Davis, J.M., and D.J. Svendsgaard. "U-Shaped Dose-Response Curves: Their Occurrence and Implications for Risk Assessment," *J. Toxicol. Environ. Health*, 30:71 (1990).
2. Holson, J.F., T.B. Gaines, C.J. Nelson, J.B. LaBorde, D.W. Gaylor, D.M. Sheehan, and J.F. Young. "Developmental Toxicity of 2,4,5-Trichlorophenoxyacetic Acid (2,4,5-T)," *Fund. Appl. Toxicol.*, 19:286 (1992).
3. Gaylor, D.W., J.J. Chen, and D.M. Sheehan. "Uncertainty in Cancer Risk Estimates," *Risk Analysis*, 13:149 (1993).
4. Snedecor, G.W., and W.G. Cochran. *Statistical Methods*, 6th ed., (Ames, IA: The Iowa State University Press, 1967), Chap. 6.
5. Beyer, W.H. *Handbook of Tables for Probability and Statistics*, 2nd ed., (Cleveland, OH: The Chemical Rubber Company, 1968), p. 304.
6. Gold, L.S., C.B. Sawyer, R. McGaw, G.M. Buckman, M. de Veciana, R. Levinson, N.K. Hooper, W.R. Havendor, L. Bernstein, R. Peto, M.C. Pike, and B.N. Ames. "A Carcinogenic Potency Database of the Standardized Results of Animal Bioassays," *Environ. Health Perspectives*, 58:9 (1984).
7. Downs, T. "Quantitative Dose-Response Models," *Biological Effects of Low Level Exposures to Chemicals and Radiation*, Vol. 1, (Chelsea, MI: Lewis Publishers, 1992), Chap. 7.
8. Morris, J.J., T.W. Roberts, and T.D. Luckey. Effects of Low Level X Radiation upon Growth of Mice, unpublished report, 1963.

9. Luckey, T.D. *Hormesis with Ionizing Radiation*, (Boca Raton, FL: CRC Press, Inc., 1980), p. 3.
10. Lorenz, E. "Some Biologic Effects of Long-Continued Radiation," *Amer. J. Roentgenol. Radium Ther. Nucl. Med.*, 63:176 (1950).

CHAPTER 7

Responses in Humans to Low Level Exposures

Ralph R. Cook, The Dow Corning Corporation,
Midland, Michigan

Epidemiology is considered by many to be a blunt instrument. As an observational science, it has a lower signal to noise ratio than do most of the experimental disciplines. In part this is because epidemiologists must deal with all the vagaries of the natural experiment. Very few of the variables are under their active direction and even with the best of conditions, the randomized clinical trial, it is impossible to control or adjust for all the technical biases that might influence the results.

As a consequence, it is often frustrating to make meaningful interpretations of data even when, in truth, there is a doubling of risk. BELLE demands that data be evaluated not to document increases in disease of 100% or more, but rather to find decreases in the order of 10% to 50%. Some epidemiologists equate this to the quest for the holy grail, i.e., a laudable but impossible task. Therefore they choose not even to look for anything but increases in effects. For them, the human biology associated with environmental exposures basically is a one-tailed statistical exercise.

That's a mistake. Even in the face of selective reporting, epidemiology research offers some tantalizing evidence that low level exposures to natural and so-called "man-made toxic chemicals" may be associated with relative risks below one. Some prefer to use the euphemisms sublinear or U-shaped dose response, hormesis, or even paradoxical effects; but whatever it's called, the results of human health research, particularly when integrated with information from other disciplines, support the necessity of investigating BELLE further.

To point out the obvious, there is a need to explore this phenomenon to better understand what the actual risks to humans are at low level exposures so that society can do a better job of setting priorities.

To point out perhaps the not so obvious, the current paradigm may be masking our abilities to identify truly adverse, albeit subtle effects. For example, if one assumes that any dose is harmful, then duration is a valid measure of exposure. However, if the assumption is incorrect, then epidemiology studies that include subjects with all durations of exposure may be mixing protective and adverse effects together, negating the ability to detect either. We lose on both sides of that coin.

There is one other point that must be considered as data are evaluated. BELLE can be approached as the yin and yang of a single outcome across a dose range; for example, the increase or decrease in a specific type of cancer at high and low dose. Or it can be looked at from the perspective of trade-offs among different outcomes at a particular dose, at the impact on the organism as a whole. In the latter case, if a low level environmental exposure markedly increases the rate of some rare health event and simultaneously halves that of a common disease, the net effect on public health could be profound.

Recently some wag wrote that the shortest distance between any two mistakes is a straight line. Nature does not operate in a statistically convenient linear fashion. That is true for biologically active chemicals whether the exposures occur naturally, intentionally, or accidentally.

Hyper- and hypo-vitamintosis have been reported for both A and D. For each, the spectrum of adverse effects differ between the "too much" and the "too little"; but with both, the human body needs some intermediate dose to function properly. At suboptimal levels of beta-carotene, various cancers are increased. At higher doses there is a decrease, particularly among those sites for which smoking is a strong risk factor.[1]

There is a comparable inverse dose response in the case of selenium. In one study, the lowest point estimates of risk for breast cancer were associated with the intermediate doses.[2] The interpretation of the epidemiology data is somewhat complicated by the observation that selenium acts as an effect modifier and not as an independent risk factor. Thus, low levels of selenium may not cause cancer so much as increase susceptibility to the effects of another carcinogen agent. There also seem to be some interactions between selenium and other micronutrients.[3]

Lead is one of the ubiquitous environmental contaminants of concern here in the United States. Among other things, overexposure can produce an anemia via the impairment of heme biosynthesis, and the increased rate of red blood cell destruction. One measure of anemia is hematocrit, the percent of whole blood which is made up of erythrocytes. In a study among children in Idaho, investigators found what they called a "slight, paradoxical increase in hematocrit at low blood lead levels."[4] While the authors dismissed this as an anomalous finding, to their credit, they reported it.

We all are bathed in a sea of chemicals that we inadvertently inhale, ingest, and adsorb. In addition, almost everyone takes a number of biologically active materials intentionally. They are called pharmaceuticals. One of the most common is aspirin. At high dose, it can produce an ototoxicity and increase the risk of stroke secondary to hemorrhage. At lower dose among men it appears to reduce cardiovascular mortality even in the face of a slightly increased risk of stroke.[5] Since the trade-off between the beneficial and the adverse effects weighs heavily toward the former, many physicians take, and advise their patients to take, low doses of aspirin on a routine basis.

Among pregnant women, aspirin is likely also associated with a U-shaped risk for intracranial cerebrovascular accidents. Pregnancy-induced hypertension (PIH) is a major cause of obstetrical and perinatal morbidity and mortality. The number of women who develop PIH is significantly lower among those treated with aspirin.[6] In other words, no aspirin and the risk of hemorrhagic stroke is high because of the elevated blood pressure. Adequate treatment and the blood pressure is controlled, the risk of stroke reduced. Too much aspirin, and the risk of a cerebrovascular bleeding increases again, even in the face of controlled blood pressure.

Alderman et al. have shown a similar U-shaped dose response associated with treatment-induced blood pressure reduction and the risk of heart attacks.[7] Both large and small reductions in diastolic blood pressure were related to a higher incidence of myocardial infarctions.

While not U-shaped, the recent favorable news about one of the most notorious drugs, thalidomide, certainly strikes many as bordering on the paradoxical. Once marketed as a sleeping pill, the drug caused thousands of birth defects. It now appears thalidomide is a potent immuno-suppressive that lacks the side effects of steroids. It is being touted as a treatment for rheumatoid arthritis, photodermatitis, leprosy, discoid lupus erythematosus, and even graft-versus-host disease, the most common cause of death following bone marrow transplantation.[8]

One of the oldest medicines known to man and a common constituent in many current over-the-counter pharmaceuticals is ethyl alcohol. In addition to having an effect on mood, at low dose elevating and at high dose depressing, those who have a drink or two a day have a lesser risk than teetotalers of suffering the most common type of stroke. In addition, they have a fraction of the risk of heavy drinkers.[9] Unfortunately, women who indulge moderately apparently also have an increased risk for breast cancer.[10]

For some reason, we all tend to be relatively blasé about exposures to natural chemicals, and intuitively accept the risk benefit inherent with medications. The same does not hold for occupational exposures. Yet many studies of chemical workers demonstrate the so-called "healthy worker effect," the more favorable health experience of the employed. This finding is usually dismissed as a function of selection—either selection of the healthy into the work force initially or selection of the susceptibles out, subsequently. To adjust for this "bias," some have even suggested that all risk estimates above 0.8 or 0.9 should be considered evidence of adverse effect![11]

In part to explore this potential bias, Ott evaluated the mortality experience of a group of chemical workers and a comparison group of employees of nonchemical industries. The cohorts were composed of ever-employed, so the selection dynamics in were likely similar. Once in the cohort, all were followed, so selection out was not an issue. The chemical workers had a more favorable mortality experience.[12]

The employees studied by Ott had been exposed to a variety of industrial materials, suggesting one or more may have had a biological effect at the levels of exposure in the workplace. If that took place, then his work further suggests that in aggregate those producing beneficial effects outweighed those leading to adverse outcomes. Others have also noted a lesser mortality among groups with heterogeneous exposures such as aerial pesticide applicators and farmers.[13-14] In each case, there are alternative explanations for the favorable findings, alternative explanations that tend to get accepted without rigorous testing because the most obvious interpretation doesn't fit with our current paradigm. It seems to lack biological plausibility. But does it?

An improved mortality experience was found among industrial workers exposed to phenol in a study conducted by the National Cancer Institute. The investigators argued that because phenol has the ability to interfere with the generation of oxidants in experimental systems, their findings may have biologic plausibility.[15]

There is some evidence that exposure to substances that cause an inflammatory response in the lung, such as bacterial endotoxins, also lead to an increase in the production of other substances such as tumor necrosis factor. This may be the explanation for the observations that people exposed to dusts containing bacterial endotoxins have a lower lung cancer incidence than those who are unexposed.[16]

The list goes on: Ethylene oxide, chlorine, arsenic, and even dioxin. Dioxin — 2,3,7,8 tetrachlorodibenzo-p-dioxin or TCDD — is considered one of the most toxic chemicals known to man. To interpret the epidemiology data properly, it is necessary to review the 1978 animal study that even to this day is considered the definitive work on dioxin.[17] Rather than present the tables as they appear, they have been reformatted, aggregating tumors by organ.

Table 7.1 provides selected data on female rats as both the absolute number of tumors by organ and the respective rates. Parenthetically, the current approach to using animal toxicology research as a predictive tool is to assume that an increase in tumors of one organ means that there will be increased frequency of tumors overall, even if there isn't. It further assumes that such an overall increase in benign and malignant tumors in animals will also occur in humans, and these tumors will be malignant. In this context, the sevenfold increase in liver tumors in the high dioxin exposure group compared to the controls is the finding that has driven most of the policies of protection, both here and abroad.

Note that this type rat has quite a high spontaneous tumor frequency, close to three tumors per animal. Note also the U-shaped dose response for total tumors. The lowest frequency was not in the controls, but rather in the low dose group. Interestingly, even in the face of a U-shaped dose response,

Table 7.1. Tumor Frequency in Female Rats Exposed to 2,3,7,8-TCDD, Rate per 100
        Animals

| Dose, µg/kg/day | 0 | | 0.001 | | 0.01 | | 0.1 | |
| Number of Rats | 86 | | 50 | | 50 | | 49 | |
	Number	Rate	Number	Rate	Number	Rate	Number	Rate
Total Tumors	230	267.4	96	192.0	102	204.0	120	244.9
Liver	9	10.5	3	6.0	20	40.0	34	69.4
Pulmonary	0	0.0	0	0.0	1	2.0	7	14.3
Ovary	3	3.5	1	2.0	1	2.0	0	0.0
Uterus	36	41.9	14	28.0	14	28.0	11	22.4
Cervix/Vagina	2	2.3	0	0.0	1	2.0	0	0.0
Mammary	81	94.2	39	78.0	40	80.0	24	49.0
Pituitary	49	57.0	18	36.0	14	28.0	14	28.6
Pancreas	5	5.8	4	8.0	1	2.0	1	2.0
Adrenal	16	18.6	8	16.0	3	6.0	8	16.3

Adapted from R.J. Kociba et al., *Toxicology & Applied Pharmacology*, 1978.

the frequency of total tumors at every dose was lower than it was for the
controls. Even for liver, there was a U-shaped dose response, albeit it was
less dramatic. For ovary, uterus, cervix/vagina, and mammary tumors there
was a downward trend, close to a reduction of 50% for uterus and mammary, but no upward swing even at the higher dose levels.

In male rats there was also U-shaped dose response for total tumors, and
the frequency also was lower than the controls at all three levels of exposure
(Table 7.2). The overall consistency of results between male and female rats
is intriguing.

Reviewing the reformatted data, the most dramatic decreases occurred in
the female primary and secondary organs of reproduction, particularly the
mammary gland. Although by convention toxicologists count both benign
and malignant tumors, epidemiologists tend to be more interested in malignant tumors. Using data kindly provided by the principal investigator of the
toxicology study, a table was constructed of the incidence of malignant
mammary tumors (Table 7.3).[18] The difference between the incidence rate in

Table 7.2. Tumor Frequency in Male Rats Exposed to 2,3,7,8-TCDD, Rate per 100
        Animals

| Dose, µg/kg/day | 0 | | 0.001 | | 0.01 | | 0.1 | |
| Number of Rats | 85 | | 50 | | 50 | | 50 | |
	Number	Rate	Number	Rate	Number	Rate	Number	Rate
Total Tumors	138	162.4	40	80.0	49	98.0	60	120.0
Liver	8	9.4	0	0.0	3	6.0	3	6.0
Pulmonary	2	2.4	0	0.0	0	0.0	2	4.0
Testes	2	2.4	2	4.0	0	0.0	0	0.0
Prostate	0	0.0	1	2.0	0	0.0	0	0.0
Mammary	2	2.4	0	0.0	0	0.0	1	2.0
Pituitary	29	34.1	6	12.0	11	22.0	13	26.0
Pancreas	29	34.1	10	20.0	8	16.0	5	10.0
Adrenal	28	32.9	6	12.0	12	24.0	9	18.0

Adapted from R.J. Kociba et al., *Toxicology & Applied Pharmacology*, 1978.

Table 7.3. Incidence of Malignant Mammary Tumors in 2,3,7,8-TCDD Exposed Female Rats

Dose, μg/kg/day	0	0.001	0.01	0.1
Number of Rats	86	50	50	49
Adenocarcinoma without metastisis	4	4	3	0
Adenocarcinoma with metastisis to lung	0	0	1	0
Adenocarcinoma with metastisis to kidney	1	0	0	0
Adenofibrocarcinoma without metastisis	3	0	2	0
Total Malignant Mammary Tumors	8	4	2	0
Incidence Rate	0.09	0.08	0.12	0.00

Adapted from R. J. Kociba et al., Results of a Two Year Chronic Toxicity and Oncogenicity Study of 2,3,7,8-Tetrachlorodibenzo-p-dioxin (TCDD) in Rats, 1977 (unpublished).

the controls and in the high dose group was statistically significant, p < 0.025.

Subsequently, Gierthy and associates showed that TCDD suppresses the postconfluent cell accumulation in estrogen-dependent MCF-7 human breast tumor cell cultures. In other words, they reported reduced formation of the multicellular foci which are characteristic of cancer cell growth in vitro.[19-20]

There was also some confirmatory information coming out of the observational research. In an epidemiology study of a large group of employees with occupational exposures to dioxin that compared the mortality experience of those with and without chloracne, a unique dermatologic condition indicative of excessive exposure to dioxin, those with chloracne appeared to have less cancer than those who did not have the condition — about 30% to 40% less.[21] However, the chloracne group was relatively small and therefore the study had low statistical power.

In 1991, National Institute for Occupational Safety and Health (NIOSH) epidemiologists published a study of 5,172 workers at 12 plants in the United States involved in the production of chemicals contaminated with TCDD. Their work did not seem to support BELLE.[22] Table 7.4 was derived from their research report and does not appear in the published paper.[23] In their journal article, the NIOSH investigators reported that the high exposure group had a significant elevation in mortality for soft tissue sarcomas — a condition of a priori concern — respiratory cancer, and all cancers combined. The latter two were what they referred to as a posteriori findings.

Parenthetically, the reason they did the study to begin with was because of a cluster of soft tissue sarcoma cases (STS) in Plants 8 and 9. It is apparent that they did not censor these sentinel cases in their analysis, so their finding of an increase in STS mortality was more a self-fulfilling prophecy than an independent test of hypothesis. Eliminating the two plants previously studied by others, they observed no mortality due to STS.

Their observations relative to lung cancer and total cancer were new; however, they were not able to address one of the major technical biases

Table 7.4. Selected Cancer Mortality Among U.S. Workers Exposed to 2,3,7,8-TCDD, Thru 1987

Plant Location	Number of Employees	STS		Lung Cancer		Total Cancer	
		Obs	SMR	Obs	SMR	Obs	SMR
1 Newark, NJ	443	0		7	72	32	115
2 Verona, MO	97	0		1	155	2	111
3 Jacksonville, AR	695	0		4	107	10	87
4 Kansas City, KS	360	0		3	101	7	75
5 Portland, OR	114	0		3	166	7	141
6 St. Louis, MO	137	0		2	104	7	129
7 Sauget, IL	96	0		3	238	3	86
8 Nitro, WV	452	2	1516[a]	15	144	35	118
9 Midland, MI	2092	2	384	28	78	105	102
10 Niagara Falls, NY	265	0		13	214[a]	30	181[a]
11 Clifton, NJ	163	0		5	239	11	193
12 Ambler, PA	258	0		5	125	16	141
TOTAL	5172	4	338	89	111	265	115[a]

[a]Statistically significant at p<0.05.
Adapted from M.A. Fingerhut et al., NIOSH Final Report, 1990.

that plague epidemiologists: confounding. They controlled for neither smoking nor alternative occupational exposures nor, in view of the fact that they identified some deaths attributed to mesotheliomas, for the synergistic effects of combined smoking and asbestos exposure. Looking at the heterogeneity of results among the total cancer SMR column, one could postulate that something other than TCDD contributed to their findings and thus masked BELLE.

In any case, none of this work on male employees addresses the issue of the potential impact of TCDD on females, in particular on human breast cancer. For that we have to turn to the reports coming out of Seveso.[24]

In 1976, an industrial accident occurred at Seveso in northern Italy that spewed dioxin over a wide area of the surrounding countryside. The various groups that were exposed have been extensively studied over the years, as has a regional control. The exposed population was divided into three separate groups based upon soil contamination. Those in Zone A were in the area of highest contamination and were evacuated within a couple of weeks of the accident. They probably received relatively acute high level exposures. Those in Zone B had some restricted access and likely had lower exposures, but for a longer period of time. Certain limitations relative to the growing of food crops were placed on the residents of Zone R, the area of lowest contamination. Since all three groups were formed as a result of the accident, there was no selection bias, i.e., this was a study in which the healthy worker effect did not potentially confound the results.

There were 556 people in Zone A; 3,920 in Zone B, 26,227 in Zone R; and 167,391 in the regional comparison group. Table 7.5 provides selected information from a 10-year follow-up. The All Causes category was chosen for perspective; All Cancers, because the All Tumors category was decreased in the animal study; Liver Cancer, because this was a key out-

Table 7.5. **Selected Mortality Among Residents of Three Zones of TCDD Contamination, Seveso, 1976–1986, Relative Risk and (95% Confidence Intervals)**

Cause of Death	Zone A N = 556	Zone B N = 3,920	Zone R N = 26,227	Zone R Obs vs Exp
All Causes				
Males	0.86 (0.5–1.4)	0.98 (0.8–1.2)	0.97 (0.9–1.0)	874 vs 901.0
Females	1.14 (0.6–2.1)	0.76 (0.6–1.0)	1.02 (0.9–1.1)	488 vs 478.4
All Cancers				
Males	0.46 (0.1–1.4)	1.19 (0.9–1.6)	0.86 (0.8–1.0)	270 vs 313.9
Females		0.6 (0.4–1.0)	0.87 (0.7–1.0)	159 vs 182.8
Liver Cancer				
Males			0.4 (0.2–0.8)	7 vs 17.8
Females			0.43 (0.1–1.4)	3 vs 7.0
Lung Cancer				
Males		1.45 (0.9–2.3)	0.78 (0.6–1.0)	77 vs 98.7
Females			0.8 (0.4–1.5)	10 vs 12.5
Breast Cancer				
Females		0.87 (0.4–2.1)	0.64 (0.4–0.9)	28 vs 43.8

Adapted from R.A. Bertazzi et al., *American Journal of Epidemiology*, 1989.

come in the toxicology research; Lung Cancer, because this also was elevated in the animal study and is a cancer of major public health significance; and Breast Cancer, because of the interesting results in multiple experimental studies. The presentation of data has been further restricted to those cells in which there were five or more observed or expected deaths. The 95% Confidence Intervals are provided in parentheses. To get some idea of the numbers involved, the last column presents the observed and expected deaths for Zone R.

Reviewing the results for Zone R, it is obvious that nothing of importance seems to have occurred relative to All Causes mortality; however, for both males and females there is a marginally significant decrease in All Cancers mortality. In view of the results in the animal study, the relative risks for both Liver Cancer and Lung Cancer are surprising. In the animal work, the frequency of tumors in both of these organs was elevated; at least it was elevated in the high dose group. In Seveso, just the opposite seems to have occurred. Among other things, the differences may be related to the species or dose level. The results for Breast Cancer are consistent with both the original toxicology study and the tissue culture research. All of this suggests something very real going on for breast cancer.

The study summarizes the mortality experience only up through 1986, about 10 years after the accident. If dioxin is an initiator, too short a time may have passed to get meaningful results. On the other hand, the results are consistent with dioxin being either preventive or therapeutic for breast cancer, and possibly for other tumors.

At the present time, whether dioxin does or does not cause cancer in humans remains controversial. If one presumes that the toxicology study is predictive and exposures are related to a sevenfold increase in certain rare tumors and simultaneously with a halving in common tumors, what are the public health implications?

In 1985, a total of 988,682 females died in the United States. Of this group, 1,029 died because of liver cancer, 17,511 of uterine or ovarian cancer; and 40,093 of breast cancer.[25] The combined total of the latter two categories was 57,604. This would suggest that for every increase in liver cancer, there would be a concomitant decrease of about 3 to 25 cancers for the other sites. From a public health perspective that seems like a pretty reasonable trade-off. Of course many would balk at the idea of causing cancer to prevent cancer. For those with such concerns, the data also suggest something more palatable. At the low doses in the animal toxicology study, there were not only fewer mammary, uterine, and ovarian tumors, there were also fewer liver tumors. And in Seveso, even significant environmental contamination was associated with a decrease in both liver and breast cancer.

In conclusion, as science becomes more sophisticated, we need to move from the paradigm of linear dose response and explore alternative models. In the process, we might also be better served by shifting from the philosophy of "above all, do no harm" to one of "do more good than harm." The key is data, especially for the latter. We must find ways of driving all segments of our society, even those with apparently different vested interests, toward the collection of ever better data. In my mind, the current paradigm does not support that. BELLE does.

REFERENCES

1. Smith, A.H., and K.D. Waller, "Serum Beta-Carotene in Persons with Cancer and Their Immediate Families," *Am. J. Epidemiol.*, 133:661 (1991).
2. Hunter, D.J., J.S. Morris, M.J. Stampfer, G.A. Colditz, F.E. Speizer, and W.C. Willett. "A Prospective Study of Selenium Status and Breast Cancer Risk," *J. Am. Med. Assoc.*, 264:1128 (1990).
3. Clark, L.C., "The Epidemiology of Selenium and Cancer," *Fed. Proceedings*, 44:2584 (1985).
4. Schwartz, J., P.J. Landrigan, E.L. Baker, W.A. Orenstein, and I.H. von Lindern. *Am. J. Public Health*, 80:165 (1990).

5. Buring, J.E., and C.H. Hennekens, "Prevention of Cardiovascular Disease: Risks and Benefits of Aspirin," *J. Gen. Internal Med.*, 5 (5 Supplement):S54 (1990).

6. Schiff, E., E. Peleg, M. Goldenberg, T. Rosenthal, E. Ruppin, M. Tamarkin, G. Barkai, G. Ben-Baruch, I. Yahal, J. Blankstein, B. Goldman, and S. Mashiach. "The Use of Aspirin to Prevent Pregnancy-Induced Hypertension and Lower the Ratio of Thromboxane A2 to Prostacyclin in Relatively High Risk Pregnancies," *New England J. Med.*, 321:351 (1989).

7. Alderman, M.H., W.L. Ooi, S. Madhavan, and H. Cohen. "Treatment-Induced Blood Pressure Reduction and the Risk of Myocardial Infarction," *J. Am. Med. Assoc.*, 262:920 (1989).

8. Randall, T., "Thalidomide's Back in the News But in More Favorable Circumstances," *J. Am. Med. Assoc.*, 263:1467 (1990).

9. Klatsky, A.L., M.A. Armstrong, and G.D. Friedman. "Alcohol and Mortality," *Annal. Internal Med.*, 117:646 (1992).

10. Longnecker, M.P., J.A. Berlin, M.J. Orza, and T.C. Chalmers. "A Meta-Analysis of Alcohol Consumption in Relation to Risk of Breast Cancer," *J. Am. Med. Assoc.*, 260:652 (1988).

11. Goldsmith, J.R., "What Do We Expect from an Occupational Cohort?" *J. Occup. Med.*, 17:126 (1975).

12. Ott, M.G., *Effects of Selection and Confounding on Mortality in an Occupational Cohort*, (Ann Arbor, MI: University Microfilms International, 1982), Chapters IV and V.

13. Stark, A.D., H. Chang, E.F. Fitzgerald, K. Riccardi, and R.R. Stone. "A Retrospective Cohort Study of Cancer Incidence Among New York State Farm Bureau Members," *Arch. Environ. Health*, 45:155 (1990).

14. Cantor, K.P., and C.F. Booze. "Mortality Among Aerial Pesticide Applicators and Flight Instructors," *Arch. Environ. Health*, 45:295 (1990).

15. Dosemeci, M., A. Blair, P.A. Stewart, J. Chandler, and M.A. Trush. "Mortality Among Industrial Workers Exposed to Phenol," *Epidemiology*, 2:188 (1991).

16. Rylander, R., "Environmental Exposures with Decreased Risks for Lung Cancer?," *Inter. J. Epidemiol.*, 19, Supplement 1:S67 (1990).

17. Kociba, R.J., D.G. Keyes, J.E. Beyer, R.M. Carreon, C.E. Wade, D.A. Dittenber, R.P. Kalnins, L.E. Frauson, C.N. Park, S.D. Barnard, R.A. Hummel, and C.G. Humiston. "Results of a Two-Year Chronic Toxicity and Oncogenicity Study of 2,3,7,8-Tetrachlorodibenzo-p-dioxin in Rats," *Toxicol. Appl. Pharmacol.*, 46:279 (1978).

18. Kociba, R.J., D.G. Keyes, J.E. Beyer, R.M. Carreon, C.E. Wade, D.A. Dittenber, R.P. Kalnins, L.E. Frauson, C.N. Park, S.D. Barnard, R.A. Hummel, and C.G. Humiston. "Results of a Two-Year Chronic Toxicity and Oncogenicity Study of 2,3,7,8-Tetrachlorodibenzo-p-dioxin (TCDD) in Rats," unpublished data, 1977.

19. Gierthy, J.F., D.W. Lincoln, M.B. Gillespie, J.I. Seeger, H.L. Martinez, H.W. Dickerman, and S.A. Kumar. "Suppression of Estrogen-Regulated Extracellular Tissue Plasminogen Activator Activity of MCF-7 Cells by 2,3,7,8-Tetrachlorodibenzo-p-dioxin," *Cancer Res.*, 47:6198 (1987).

20. Gierthy, J.F., and D.W. Lincoln. "Inhibition of Postconfluent Focus Production in Cultures of MCF-7 Human Breast Cancer Cells by 2,3,7,8-

Tetrachlorodibenzo-p-dioxin," *Breast Cancer Research and Treatment*, 12:227 (1988).

21. Bond, G.G., R.R. Cook, F.E. Brenner, and E.A. McLaren. "Evaluation of Mortality Patterns Among Chemical Worker with Chloracne," *Chemosphere*, 16:2117 (1987).

22. Fingerhut, M.A., W.E. Halperin, D.A. Marlow, L.A. Piacitelli, P.A. Honchar, M.H. Sweeney, A.L. Griefe, P.A. Dill, K. Steenland, and A.J. Suruda. "Cancer Mortality in Workers Exposed to 2,3,7,8-Tetrachlorodibenzo-p-dioxin," *New England J. Med.*, 324:212 (1991).

23. Fingerhut, M.A., W.E. Halperin, D.A. Marlow, L.A. Piacitelli, P.A. Honchar, M.H. Sweeney, A.L. Griefe, P.A. Dill, K. Steenland, and A.J. Suruda. "Mortality Among U.S. Workers Employed in the Production of Chemicals Contaminated with 2,3,7,8-Tetrachlorodibenzo-p-dioxin (TCDD)," *NIOSH Final Report PB91125971*, National Technical Information Service, Springfield, VA, 1990.

24. Bertazzi, R.A., C. Aocchetti, A.C. Pesatori, S. Guercilena, M. Sanarico, and L. Radice. "Ten-Year Mortality Study of the Population Involved in the Seveso Incident in 1976," *Am. J. Epidemiol.*, 129:1187 (1989).

25. National Center for Health Statistics, *Vital Statistics of the United States, 1985*, II, Mortality, Part A, DHHS Publication (PHS) 88–1101, U.S. Government Printing Office, Washington, DC, 1988, p. 232.

CHAPTER 8

Cellular and Molecular Foundations of Hormetic Mechanisms

Harihara M. Mehendale, Division of Pharmacology and Toxicology, College of Pharmacy and Health Sciences, Northeast Louisiana University, Monroe, Louisiana

INTRODUCTION

All forms of life struggle to survive the challenges encountered in the environment. Toxic injury by natural or man-made hazardous chemicals or physical agents is one such challenge to living organisms. Since the basic unit of life is the cell, cellular compensatory responses to such injurious episodes serve as the first line of defense. Cellular responses include reinforcement of cytoprotective mechanisms intended to protect the cell. If these mechanisms are overcome by the toxic insult, another level of cellular and molecular protective mechanisms is unleashed. These mechanisms may either be activated, or may be induced by a previous episode of exposure to a toxic agent. These protective events might be collectively referred to as hormetic mechanisms. Exposure to high levels of chemicals results in either suppression or a significant delay in the expression of these hormetic mechanisms. Therefore, information on potential for adverse health effects from exposure to chemicals derived from exposure to high levels may not be very useful in assessing the risk to public health in the best case, or may be misleading, in the worst case. In recent years, much information has become available on the cellular and molecular basis of hormesis through studies on the biological effects of low levels of exposure to either single or combinations of chemicals.

What about simultaneous or sequential exposure to more than one chemical? From a perspective of public health, a major toxicological issue is the possibility of unusual toxicity due to interaction of two or more toxic chemicals at individually harmless levels upon environmental or occupational exposures. While some laboratory models exist for such interactions involving two chemicals, progress in this area has suffered for want of models where the two interactants are individually nontoxic. Toxicities resulting from exposure to more than two chemicals at individually nontoxic doses are of greater interest since this exposure scenario is most common. One such model is available, where prior exposure to nontoxic levels of the pesticide Kepone® (chlordecone) results in a 67-fold amplification of CCl_4 lethality in rats (Table 8.1). The mechanism of this remarkable interactive toxicity is of interest in the assessment of risk from exposure to combinations of chemicals.

CCl_4 Toxicity Amplified by Dietary Exposure to Chlordecone

Prior exposure to nontoxic level of chlordecone (10 ppm in diet for 15 days) results in a marked amplification of CCl_4 hepatotoxicity[1-3] and lethality.[3-5] Neither the close structural analogs of chlordecone, mirex, and photomirex, nor phenobarbital (Figure 8.1), exhibit this property.[2,3] Others[6,7] have demonstrated the capacity of chlordecone to potentiate $CHCl_3$ hepatotoxicity in mice. These observations have been extended to demonstrate that in addition to the hepatotoxic effects, lethal effect of $CHCl_3$ is also potentiated by exposure to 10 ppm dietary chlordecone[8] (Table 8.2), and that this is also associated with suppressed repair of the liver tissue.[9] Chlordecone also potentiates the hepatotoxicity and lethality of $BrCCl_3$.[10,11] While the toxicity of these closely related halomethanes is potentiated by such low levels of chlordecone (Figure 8.2), the toxicity of structurally and mechanistically dissimilar compounds (Figure 8.3, Table 8.3) is not potentiated,[12] except after exposure to high levels of chlordecone.[13] This remarkable propensity to potentiate halomethane hepatotoxicity does not appear to be related to chlordecone-induced cytochrome P-450 or associated enzymes,[2,5,14] enhanced bioactivation of CCl_4,[12,15-18] increased lipid peroxidation,[2,12,13] or decreased glutathione.[19] Several candidate mechanisms were found to be

Table 8.1. Amplification of Lethal Effects of Several Halomethanes by Dietary Exposure of Rats to Subtoxic Contaminants

Dietary Pretreatment	Halomethane	48 hr LD$_{50}$ mL/kg	Increase in Toxicity -fold
Female Rats			
Control	CCl_4	1.25	–
Chlordecone (10 ppm)	CCl_4	0.048[a]	26
Male Rats			
Control	CCl_4	2.8	–
Chlordecone (10 ppm)	CCl_4	0.042[a]	67
Phenobarbital (225 ppm)	CCl_4	1.7	1.6[b]
Control	$BrCCl_3$	0.119	–
Chlordecone (10 ppm)	$BrCCl_3$	0.027[a]	4.5

[a]Highly significant compared to the respective solvent control.
[b]Not significant at $P \leq 0.05$. Adapted from Reference #1.

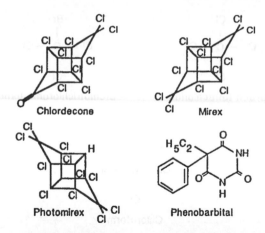

Figure 8.1. Structures of chlordecone, mirex, photomirex, and phenobarbital. Chlordecone (Kepone®) amplifies the toxicity of halomethanes closely related to CCl_4. Despite being close structural analogs of chlordecone, mirex and photomirex do not possess this propensity. Phenobarbital, a commonly employed drug in interaction studies at high doses, does increase liver injury of CCl_4, but this enhanced liver injury is inconsequential to animal survival, since these animals are able to recover from liver injury due to greatly stimulated liver tissue repair.

inadequate, and additional experiments revealed a novel mechanism involving precipitous decline in cellular energy and suppressed tissue repair (Table 8.4).

Mechanism of Chlordecone-Amplified Toxicity of CCl_4

These findings led to some very basic studies concerning the progression of the hepatotoxicity during a time-course following CCl_4 administration to either normal or chlordecone-pretreated rats. The histochemical and histomorphometric experiments revealed that suppressed hepatocellular regeneration and tissue repair[1,2,20,21] might explain the remarkable amplification of CCl_4 toxicity by prior exposure to chlordecone. Similar time-course studies on Ca^{2+} levels in the liver mitochondria, microsomes, and cytosol frac-

Table 8.2. Amplification of Lethal Effects of Halomethanes by Dietary Exposure of Mice to Subtoxic Contaminants

Dietary Pretreatment	Halomethane	48 hr LD_{50} mL/kg	Increase in Toxicity -fold
Male Rats			
Control	$CHCl_3$	0.067	–
Chlordecone (10 ppm)	$CHCl_3$	0.16[a]	4.2
Mirex (10 ppm)	$CHCl_3$	0.70	No change
Phenobarbital (225 ppm)	$CHCl_3$	0.70	No change

[a]Significantly different at $P \leq 0.05$. Adapted from Reference #1.

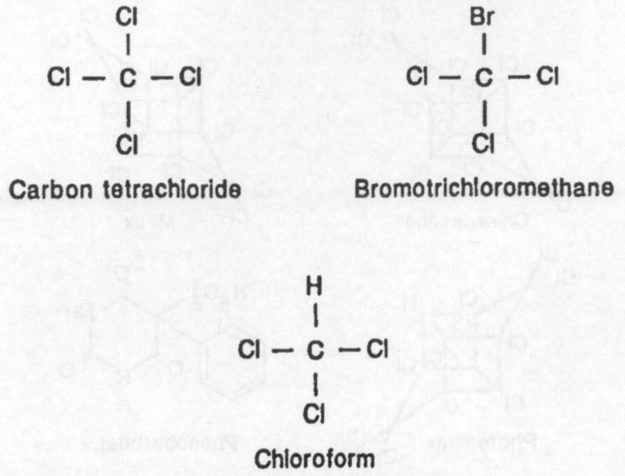

$$
\begin{array}{ccc}
& Cl & \\
& | & \\
Cl - & C & - Cl \\
& | & \\
& Cl &
\end{array}
\qquad
\begin{array}{ccc}
& Br & \\
& | & \\
Cl - & C & - Cl \\
& | & \\
& Cl &
\end{array}
$$

Carbon tetrachloride   Bromotrichloromethane

$$
\begin{array}{ccc}
& H & \\
& | & \\
Cl - & C & - Cl \\
& | & \\
& Cl &
\end{array}
$$

Chloroform

Figure 8.2. Structures of carbon tetrachloride, bromotrichloromethane and chloroform as examples of halomethane solvents. Hepatotoxicity and lethality of these solvents are remarkably amplified by chlordecone.

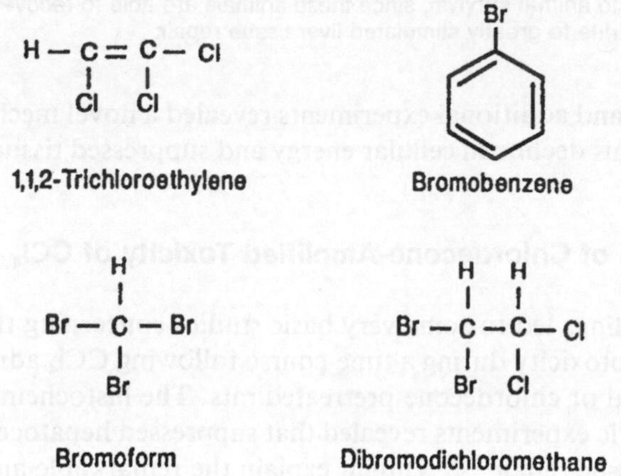

$$
\begin{array}{c}
H - C = C - Cl \\
\;\; | \quad\; | \\
\;\; Cl \;\;\; Cl
\end{array}
$$

1,1,2-Trichloroethylene   Bromobenzene

$$
\begin{array}{c}
H \\
| \\
Br - C - Br \\
| \\
Br
\end{array}
\qquad
\begin{array}{c}
H \;\; H \\
| \;\;\; | \\
Br - C - C - Cl \\
| \;\;\; | \\
Br \;\; Br
\end{array}
$$

Bromoform   Dibromodichloromethane

Figure 8.3. Structure of 1,1,2-trichloroethylene, bromobenzene, bromoform, and dibromodichloromethane. Toxicity of these chemicals is not potentiated by prior dietary exposure to 10 ppm chlordecone.

Table 8.3. Specificity of Potentiation of Halomethane Toxicity by Chlordecone

Compound	Potentiation?	Reference
$CHCl_3$	yes	6, 8
CCl_4	yes	2, 3
$CBrCl_3$	yes	10,11
$CHBr_3$	no	2
CBr_4	no	2
CCl_2CHCl	no	12
Bromobenzene	no	12

Table 8.4. Specificity of Potentiation of Halomethane Toxicity by Chlordecone

Mechanism	Role in Amplification
1. Enhanced bioactivation of halomethanes.	Increased infliction of injury. Only Stage I of toxicity is increased.
2. Increased lipid peroxidation.	Not known or none.
3. Estrogenic property of chlordecone.	None
4. Increased Ca^{2+} accumulation. Precipitous glycogenolysis & loss of ATP.	Perturbed cellular biochemistry and ablation of hormetic mechanisms.
5. Suppressed hepatocellular regeneration and unabated progression to Stage II of toxicity.	Injury becomes irreversible due to ablation of the early-phase hormesis.

tions[22,23] revealed a possible association of increased Ca^{2+} accumulation and suppressed hepatocellular regeneration. Despite some reports that chlordecone interferes with Ca^{2+} uptake mechanisms in extrahepatic tissues,[24] chlordecone alone does not cause disruption of hepatocellular Ca^{2+} [25] even at toxic doses. Chlordecone + CCl_4 interaction results in remarkably increased intracellular Ca^{2+}.[22-26] Recent studies have also shown a significant activation of phosphorylase a, a finding consistent with the precipitous depletion of glycogen[21,27-31] and ATP.[27-31] The hypothesis proposed[1] for the mechanism (Table 8.4) of chlordecone-amplified toxicity of CCl_4 has received substantial experimental support.

Stimulation of Tissue Repair as a Hormetic Response to Tissue Injury

First, it became necessary to investigate the mechanism for *why an ordinarily nontoxic dose of CCl_4 is nontoxic?* Figure 8.4 illustrates the mechanism of recovery from limited liver injury observed after the administration of a low dose of CCl_4 alone. Within 6 hr after the administration of a low dose of CCl_4, limited hepatocellular necrosis accompanied by ballooned cells and steatosis inflicted by the same widely accepted mechanisms of CCl_4 bioactivation followed by lipid peroxidation occurs. Simultaneously, the liver tissue responds by stimulating hepatocellular regeneration.[20,21] Most interestingly, one burst of hepatocellular division is evident at 6 hr, even though liver injury evident as centrilobular necrosis only begins to manifest itself at that time. Although the molecular events responsible for the stimulation of hepatocellular division have not been explored, glycogen, the principal form of stored hepatic energy resource, is mobilized prior to cell division.[20,21] Glycogen levels are restored after cell division has been adequately stimulated.[20,21] The limited hepatocellular necrosis enters the progressive phase between 6 and 12 hr,[20,21,32,33] while the hepatocellular regeneration and tissue healing processes continue. By 24 hr, no significant liver injury is evident, and by 36 hr the liver appears completely normal. These observations indicate that stimulation of hepatocellular regeneration is a protective response of the liver, occurs very early after the administration at low dose of CCl_4, and leads to replacement of dead cells, thereby restoring the hepatolobular architecture.[2,35]

Furthermore, this remarkable biological event results in another important protective action. It is known that newly divided liver cells are relatively resistant to toxic chemicals.[36-42] Therefore, in addition to the restoration of the hepatolobular architecture by cell division, by virtue of the relatively greater resistance of the new cells, the liver tissue is able to overcome the imminence of greater injury during the progressive phase (6 to 12 hr), preventing the spread of injury on the one hand, and speeding up the process of overall recovery through tissue healing, on the other (Figure 8.4). By 6 hr over 75% of the administered CCl_4 is eliminated in the expired air,[14] leaving less than 25% in the animal.[2] At later time points (12 hr and onward), most of the CCl_4 would have been eliminated by the animal, thereby preventing additional infliction of injury. Continued cellular regeneration during this time period and at later time points allows for complete restoration of the hepatolobular architecture during and after the progressive phase of injury.[32,33,43,44] Relative resiliency of the newly divided cells at this critical time frame, as the animal continues to exhale the remaining CCl_4, is an added critical defense mechanism easily available through cell division.

Administration of the same low dose of CCl_4 to animals maintained on food contaminated with low dose of chlordecone results in initial injury by the same mechanisms of bioactivation of CCl_4 and lipid peroxidation (Figure 8.4). The liver injury in this case is slightly greater due to approximately doubled rate of bioactivation of CCl_4 in livers of animals preexposed to chlordecone.[2,14,35] The liver injury, thus initiated, enters the progressive phase between 6 to 12 hr, and this phase is accelerated in the absence of tissue repair mechanisms.[20,21,32,33, 43,44] The highly amplified toxicity of CCl_4 is a consequence of the suppression of hepatocellular regeneration and tissue repair, otherwise ordinarily stimulated by CCl_4 (Figure 8.4).

The mechanism responsible for the abrogation of this hormetic response of stimulated cell division is of significant interest. Substantial experimental observations indicate that a lack of hepatocellular energy leads to a failure of cell division. Under conditions of increased hepatocellular injury, mobilization of hepatic glycogen is initiated in order to stimulate hepatocellular division.[21-26] Insufficient energy at a time of increased demand for cellular energy (augmented need for extrusion of extracellular Ca^{2+} from the cells, protection against free-radical mediated injury, etc.), incapacitates the hepatocytes.[27] As a result, stimulation of cell division, which normally occurs after the administration of a low dose of CCl_4, cannot occur.[31] The failure of cell division has two important implications: first, hepatolobular structure cannot be restored; second, unavailability of newly divided, relatively resistant cells predisposes the liver to a permissive continuation of liver injury during the progressive phase (6 to 12 hr and beyond).[1,2,26,35,45,46]

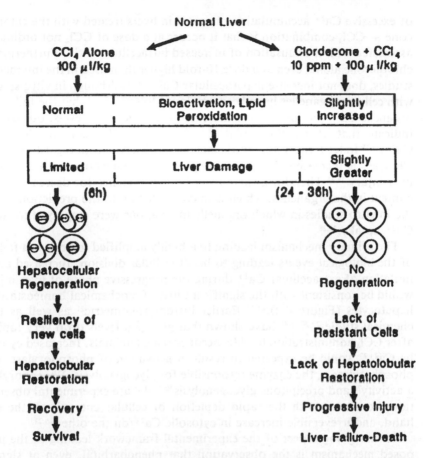

Figure 8.4. Proposed mechanism for the highly amplified interactive toxicity of chlordecone + CCl$_4$. The scheme depicts the concept of suppressed hepatocellular regeneration, simply permitting what is normally limited liver injury caused by a subtoxic dose of CCl$_4$ to progress in the absence of hepatolobular repair and healing mechanisms stimulated by the limited injury. The limited hepatotoxicity from a low dose of CCl$_4$ is normally controlled and held in check owing to the hepatocellular regeneration and hepatolobular healing. The chlordecone + CCl$_4$ combination treatment results in unabated progression of injury owing to a lack of tissue repair obtunded due to lack of cellular energy. These events lead to complete hepatic failure, culminating in animal death. Ongoing studies indicate that a very similar mechanism is responsible for the amplification of CHCl$_3$ and BrCCl$_3$ toxicity by chlordecone. Adapted from Reference #1.

Permissively progressive injury continues unabatedly as a consequence of the mitigated tissue repair mechanisms, leading to massive hepatic failure,[1,4,5,10] followed by animal death.[1,4,21-46]

The mechanism underlying a rapid and precipitous decline in cellular ATP is of considerable interest. Many studies have shown a biphasic increase in hepatocellular Ca^{2+} levels in CCl$_4$ toxicity.[23] The unusual aspect of excessive Ca^{2+} accumulation observed in livers treated with the chlordecone + CCl_4 combination is that it occurs at a dose of CCl_4 not ordinarily associated with the causation of increased intracellular Ca^{2+}. Furthermore, chlordecone alone, even at a dose 10-fold higher than used in the interaction studies, does not increase hepatocellular Ca^{2+}.[22,25] Although in vitro studies with cellular organelles have been employed to speculate that the failure of organelle Ca^{2+} pumps leads to increased cytosolic Ca^{2+} levels, our studies indicate that at no time-point do these organelles contain decreased Ca^{2+}.[23,35] Indeed, the only significant change observed with regard to organelle Ca^{2+} is increased Ca^{2+} in the organelles in association with increased liver injury.[35,45] Therefore, there is no in vivo evidence for decreased Ca^{2+} content in the organelles, which is in contradiction to the predictions from the in vitro studies in which organelle incubations were employed to study Ca^{2+} uptake.[26,28]

The primary mechanism leading to a highly amplified toxicity is a failure of the biological events leading to hepatocellular division. Increased accumulation of extracellular Ca^{2+} during the progressive phase of liver injury would be consistent with the significant loss of biochemical homeostasis in hepatocytes (Figure 8.4).[23,27] Earlier histomorphometric[21] as well as biochemical studies[29,30,35,45] have shown that glycogen levels drop very rapidly after CCl_4 administration to chlordecone-treated animals. Increased cytosolic Ca^{2+}[27] would be expected to result in activation of phosphorylase b to phosphorylase a, the enzyme responsible for glycogenolysis. Phosphorylase a activity[26] and precipitous glycogenolysis[20,21,23,26] are experimental observations consistent with the rapid depletion of cellular energy[28] on the one hand, and irreversible increase in cytosolic Ca^{2+} on the other.[26,27]

An intriguing aspect of the experimental framework leading to the proposed mechanism is the observation that phenobarbital, even at significantly higher doses (225 ppm in the diet for 15 days) does not potentiate the lethal effect of CCl_4. Although histopathological parameters of liver injury such as hepatocellular necrosis and ballooned cell response are indicative of significantly enhanced hepatotoxicity by phenobarbital, if the animals are left alone, this injury does not progress to significantly increased lethality. Hepatic microsomal cytochrome P-450 is approximately doubled by prior dietary exposure to 225 ppm PB and the bioactivation of CCl_4 is tripled,[2,14] and these indicators are consistent with the enhanced initiation of liver injury (Stage I of toxicity) measured by histopathology, elevation of serum transaminases, or by hepatic function. Nevertheless, the liver injury neither progresses in an accelerated fashion nor is irreversible, as indicated by the reversal of liver injury accompanied by animal survival.[12,4,33]

Figure 8.5 illustrates the proposed mechanism for phenobarbital-enhanced CCl_4 liver injury, which is not associated with increased lethality. Induction of hepatomicrosomal cytochrome P-450 results in approximate tripling of CCl_4 bioactivation and increased lipid peroxidation.[2,14] Enhanced liver injury is consistent with these observations (Figure 8.5). It should be

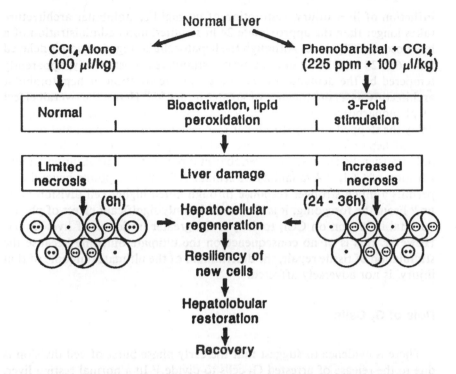

Figure 8.5. Proposed mechanism for phenobarbital-induced potentiation of CCl₄-hepatotoxicity in the absence of increased lethality. Normal liver response to a low dose-CCl₄ injury is not abrogated by phenobarbital + CCl₄ interaction. Instead, the early phase of cell division is postponed (from the normal 6 hr to 24 hr). Enhanced putative mechanisms such as increased bioactivation of CCl₄ and resultant increased lipid peroxidation are responsible for the increased infliction of Stage I injury. Because hepatocellular regeneration and tissue repair processes continue albeit a bit later than normal, these hormetic mechanisms permit tissue restoration resulting in recovery from the enhanced liver injury. This mechanism explains the remarkable recovery from phenobarbital-induced enhancement of CCl₄ liver injury. Despite a remarkably enhanced liver injury by phenobarbital, this is of no real consequence to the animal's survival because depletion of cellular energy does not occur with this interaction, which permits hormetic mechanisms to restore hepatolobular architecture, resulting in complete recovery. Adapted from Reference #48.

recalled that the liver is normally able to respond by stimulation of hepatocellular regeneration after a low dose of CCl₄ within 6 hr (Figure 8.4). While phenobarbital exposure results in greater infliction of CCl₄ injury, the ability of the liver to respond by stimulated cell division is not compromised, as evidenced by the stimulation of hepatocellular regeneration starting at 24 and continuing through 72 hr. Hepatocellular regeneration is greatly stimulated, thereby counteracting the enhanced liver injury, which leads to recovery despite the greater infliction of liver injury. In view of the greater infliction of liver injury, restoration of normal hepatolobular architecture takes longer than the approximate 24 hr required upon administration of a low dose of CCl_4 alone. Although the hepatocellular regeneration is delayed from 6 to 24 hr, when it does occur it is stimulated substantially, apparently tempered by the demand for more extensive restoration of hepatolobular architecture as a consequence of greater injury.[8,9,33] Hence the overall effect of phenobarbital-induced potentiation of CCl_4 injury is merely to slightly delay the stepped-up hepatocellular regeneration, tissue repair, and restoration of hepatolobular architecture. Interestingly, hepatic ATP levels were only transiently decreased in phenobarbital pretreated animals upon administration of CCl_4.[28] Availability of cellular ATP at time points beyond 6 hr permits a much stronger response through much higher cell division at 24 hr.[33] From the foregoing, it is clear that the only significant effect of phenobarbital induction on CCl_4 toxicity is the greater infliction of liver injury. However, this is of no consequence on the ultimate outcome because the stimulation of tissue repair, the determinant of the ultimate outcome of that injury, is not adversely affected.

Role of G₂ Cells

There is evidence to suggest that the early-phase burst of cell division is due to the release of arrested G_2 cells to divide.[47] In a normal resting liver, 4% to 6% of the hepatocytes are in G_2 phase (Figure 8.6). These cells have been shown to divide after administration of a variety of chemicals.[43] Six hr after the administration of a low dose of CCl_4, G_2 cells are released to divide (Figure 8.6). A second wave of cell division is stimulated, resulting presumably from the mitotic stimulus of endogenous growth factors such as Tissue Growth Factor-alpha (TGF-α). High dose of CCl_4, prior exposure to a nontoxic dose of chlordecone, or colchicine are known to ablate the stimulation of G_2 cell division. Moreover, the second wave of cell division resulting from the initiation of G_0 cells is also attenuated and/or delayed. Under these conditions, the toxic injury enters the progressive phase. Prior exposure to phenobarbital results in a slight delay in the early-phase burst of cell division, but then the response is much stronger (Figure 8.6).[47]

Critical Role of Stimulation of Cell Division and Tissue Repair

Table 8.5 presents a variety of experimental manipulations that permit a rigorous experimental verification of the existence and the critical role played by tissue repair in the final outcome of toxic injury. The experimental evidence for the existence of a hormetic mechanism was derived as the result of persistent efforts to understand the mechanism of chlordecone potentiation of halomethane toxicity.

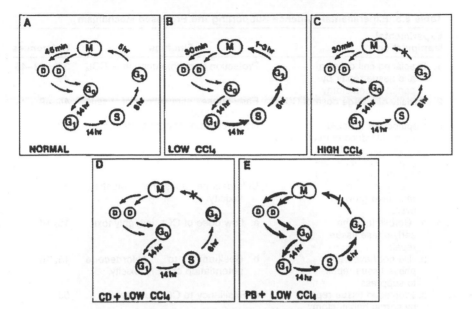

Figure 8.6. Schematic illustration of hepatic cell cycle and the involvement of G_2 cells in the toxicology of low dose CCl_4, chlordecone + CCl_4, phenobarbital + low dose CCl_4, and high dose of CCl_4. Adapted from Reference #47.

A. Normal cell cycle processes are illustrated in an idealized scheme. Normally, cell division is very quiescent in adult livers; is sufficient to replace apoptotic loss of cells and restoration of lobular structure and function. At any given time, a normal liver cell population might be ranked approximately as follows. $G_0 >>> G_1 >> S > G_2 > M$. Prereplication changes may take approximately 14 hr and then another 14 hr for S phase DNA synthesis, followed by approximately 8 hr for G_2 phase. Cells in G_2 phase will reach M phase relatively quickly.

B. After administration of a low dose of CCl_4 (e.g., 100 μL/kg i.p.), cells (thick arrow) in G_2 phase are stimulated to divide giving rise to the early-phase response (6 hr). S-phase is also known to be stimulated by a low dose of CCl_4. Furthermore, the entire cell cycle is enhanced resulting in liver cell division peaking between 36–48 hr.

C. After high dose CCl_4 (2.5 mL/kg, i.p.) the early-phase forward progression of G_2 to M phase is inhibited. The second wave of cell division, resulting from forward progress of G_1 cells is stimulated, but this response is insufficient because of progressive necrotic and degenerative events consequent to the obtunded early-phase cell division. Hepatic injury becomes progressive and a dose-dependent expression of ultimate toxicity is observed.

D. Upon exposure to dietary 10 ppm chlordecone for 15 days, the early-phase cell division response normally seen after a low dose of CCl_4 ($G_2 \rightarrow M$) is suppressed. There is evidence suggesting that this is due to insufficient cellular energy. Forward progress of $G_1 > M > S > G_2 > M$ is stimulated, but to a lesser extent. There is evidence to indicate that insufficient cellular energy obtunds $G_2 > M$ progression while slowing down the entire cell cycle.

E. Upon exposure to 225 ppm dietary phenobarbital for 15 days, the early-phase stimulation of liver cell division by a low dose of CCl_4 ($G_2 > M$) is decreased. However, the overall cell cycle time is significantly decreased as evidenced by a very substantial increase in cell division starting at 24 hr and peaking at 48 hr. These events enable the animals to overcome liver injury.

Table 8.5. Experimental Evidence Supporting the Proposed Mechanism

Experimental Manipulation	Findings	References
1. Preplaced cell division and tissue repair by partial hepatectomy.	Protection from chlordecone + CCl_4	32, 43–45, 49
2. Toxicity of a large dose of CCl_4.	Early-phase stimulation of tissue repair is ablated.	44, 49
3. Hepatocytes isolated from chlordecone treated rats incubated with CCl_4. (isolated hepatocytes do not divide in vitro)	No potentiation in contrast to in vivo.	52
4. Developing young rats have growing livers.	Chlordecone does not potentiate toxicity.	41
5. a. Gerbils lack the early-phase tissue repair	a. Low dose of CCl_4 is highly toxic.	15, 16
b. Do not have early-phase tissue repair to suppress.	b. Resilient to chlordecone potentiation of CCl_4 toxicity.	15, 16
c. Preplaced tissue repair by partial hepatectomy.	c. Resiliency to CCl_4 toxicity.	53
6. CCl_4 autoprotection.	Due to prestimulation of early-phase tissue repair by the protective dose.	54
7. a. Selective ablation of the early-phase hormesis by colchicine.	a. Prolongation of hepatotoxicity of a low dose of CCl_4 by 24 hr (until the second phase of cell division at 48 hr ensues to overcome injury).	55–57
b. Colchicine given 2 hr before the protective dose of CCl_4.	b. Abolishes CCl_4 autoprotection entirely.	57

Adapted from Reference #50.

1. Partial Hepatectomy

If the basic premise is valid that suppression of the early-phase (6 hr) stimulation of cell division and tissue repair is the mechanism of chlordecone potentiation of CCl_4 injury, then a preplacement of cell division in the liver should result in protection against the interactive toxicity of chlordecone + CCl_4. When CCl_4 was administered at 2 days after partial hepatectomy at a time of maximally stimulated hepatocellular division, a remarkable protection was observed.[48] At 7 days after partial hepatectomy when the stimulated cell division phases out, the interactive toxicity becomes fully manifested again.[48] In these studies, microsomal cytochrome P-450 content is decreased by partial hepatectomy, but remains at the decreased level even 7 days later when protection is no longer evident. Moreover, actual in vivo bioactivation, and overall disposition of $^{14}CCl_4$ is unperturbed by partial hepatectomy.[18]

2. Large Dose is Toxic Owing to Attenuated and Delayed Hormetic Response

An implication of these findings is that the toxic effect of a large dose of CCl$_4$ might be a consequence of suppressed cell division and tissue repair. When a large dose of CCl$_4$ was administered, the early-phase cell division normally stimulated by a low dose of CCl$_4$[20,21,33,44] was ablated entirely.[44,51,54] These findings indicate that the real difference between a low and a high dose of CCl$_4$ is the presence or absence of promptly stimulated cell division and tissue repair. The higher dose clearly prevents the prompt and adequate hormetic response, thus permissively allowing toxicity to progress unabatedly.

3. Chlordecone Amplification of CCl$_4$ Toxicity Does Not Occur Under in Vitro Conditions Where Cell Division and Tissue Repair Cannot Be Expressed

Yet another line of experimental validation of the critical role of suppressed cell division and tissue repair comes from in vitro incubation of hepatocytes isolated from chlordecone-pretreated rats with CCl$_4$.[52] Isolated hepatocytes do not divide under in vitro conditions. Therefore, if suppression of cell division and tissue repair ordinarily stimulated by a low dose of CCl$_4$ is the mechanism of chlordecone-amplified CCl$_4$ toxicity, one should not observe highly amplified toxicity when hepatocytes from chlordecone-treated rats are incubated with CCl$_4$ in vitro. Since prior exposure to phenobarbital is known to result in increased CCl$_4$ toxicity in vitro, incubation of hepatocytes obtained from phenobarbital-treated rats with CCl$_4$ should result in a measurable level of increased toxicity. Such experiments revealed no significant increase in cytotoxic injury in chlordecone-pretreated isolated hepatocyte incubations.[52] Cells from phenobarbital-pretreated rats exhibited highest CCl$_4$ toxicity indicating that the in vitro paradigm was working as expected. These findings are consistent with the hypothesis that suppression of hepatocellular division and tissue repair is the primary mechanism of chlordecone-potentiated CCl$_4$ toxicity, and provide substantial evidence against any significant role for chlordecone-enhanced bioactivation of CCl$_4$.[52]

4. Resiliency of Newborn and Developing Rats

Newborn and postnatally developing rats have actively growing livers. Since livers during active growth have ongoing cell division, these developing rats would be expected to be resilient during their early development. When rat pups at 2,5,20,35,45 and 60 days were tested, they were completely resilient to chlordecone-potentiation of CCl$_4$ toxicity up to 35 days of age.[40,41] At 45 days, young rats were sensitive to the interactive toxicity of chlordecone + CCl_4 and by 60 days the rats were just as sensitive as adults.[40] The hepatic microsomal cytochrome P-450 levels in the livers of 35,45 and 60-day old rats exposed to chlordecone were not different from each other, suggesting that any differences in cytochrome P-450 levels are unlikely to explain the observed differences in toxicities. Moreover, recent studies indicate that bioactivation of $^{14}CCl_4$ in 35-day old rats is not less than that observed in 60-day old rats.[40] Therefore, the resiliency of younger rats to chlordecone-potentiation of CCl_4 toxicity is more likely related to the ongoing hepatocellular regeneration during early development rather than due to differences in the bioactivation of CCl_4.

5. Gerbils Lack the Early-Phase Hormesis and Are Most Sensitive to Halomethane Toxicity

While administration of a low dose of CCl_4 to rats results in a prompt stimulation of early-phase hepatocellular regeneration at 6 hr[32,33,43,44,49] in Mongolian gerbils, this early-phase cell division is not observed.[16] The stimulation of cell division which occurs at 42 hr (analogous to the second phase of cell division occurring at 48 hr in rats) appears to be too little and too late to be of any help in overcoming liver injury.[15,16] If the early-phase cell division is critical for recovery from liver injury, then owing to a lack of this important hormetic mechanism in gerbils, they should be extremely sensitive to halomethane toxicity. When tested, gerbils were found to be approximately 35-fold more sensitive to the toxicity of CCl_4.[15] Likewise, gerbils show several-fold greater sensitivity to the lethal effects of $BrCCl_3$ and $CHCl_3$ (Tables 8.5 and 8.6). It follows that gerbils should not be susceptible to chlordecone-potentiation of CCl_4 toxicity (Table 8.6) since they lack the early-phase of hepatocellular regeneration, the target of that interaction.[16] Studies have shown that a preplacement of hepatocellular regeneration by partial hepatectomy results in significant protection against CCl_4 toxicity,[53] underscoring the importance of stimulated hepatocellular regeneration in determining the final outcome of liver injury. These studies also reveal

Table 8.6. High Sensitivity of Mongolian Gerbils to Halomethane Toxicity Contrasted with Their Resiliency to Potentiation by Exposure to Other Chemicals

| Halomethane | Normal Diet | 15-day Dietary Pretreatment | | |
		Chlordecone (10 ppm)	Phenobarbital (225 ppm)	Mirex (10 ppm)
		mL/kg		
CCl_4	80	100	100	100
	(34–186)[a]	(78–128)	(28–354)	(28–354)
$CBrCl_3$	20	20	20	16.8
	(8.6–46.5)	(16.4–24.4)	(10.4–38.4)	(9.9–28.6)
$CHCl_3$	400	565	400	400
	(208–769)	(346–923)	(268–597)	(268–597)

[a]95% Confidence intervals. Adapted from Reference #16.

another important difference between species. While rats respond by maximal stimulation of hepatocellular regeneration within 2 days after partial hepatectomy, in gerbils the maximal stimulation was many-fold lower and it occurred not before 5 days after partial hepatectomy.[53] These findings suggest that gerbils are much more sluggish in their hormetic response to a noxious challenge of a hepatotoxic chemical agent. Each of these findings points to the critical importance of the early-phase stimulation of cell division as a decisive target of inhibition in chlordecone-amplification of CCl_4 toxicity (Table 8.4). Secondly, these findings also underscore the importance of the biological hormetic response in determining the resiliency to the toxic action of halomethanes.

6. Autoprotection

CCl_4 autoprotection is a phenomenon whereby administration of a single low dose of CCl_4 24 hr prior to the administration of a killing dose of the same compound results in an abolition of the killing effect of the large dose.[54-58] The widely accepted mechanism of this phenomenon is the destruction of liver microsomal cytochrome P-450 by the protective dose such that subsequently administered large dose is insufficiently bioactivated.[59-65] Since bioactivation of CCl_4 is an obligatory step for its necrogenic action, it was suggested that massive liver injury ordinarily expected from a large dose of CCl_4 never occurs in the autoprotected animal.[34] Although this mechanism has been widely accepted, a closer examination of the evidence suggests that the mechanism was largely derived by association[58-64] rather than actual experimental evidence of substantially lower liver injury in the autoprotected animal. Additionally, several lines of evidence indicate that even after the significant destruction of cytochrome P-450, the availability of the P-450 isozyme responsible for the bioactivation of CCl_4 is not limiting.[18,49,50,54,66-69] For instance, even after a 60% decrease in the constitutive liver microsomal cytochrome P-450 by $CoCl_2$ treatment, CCl_4 toxicity was undiminished regardless of whether the rats were pretreated with chlordecone.[49] More direct evidence was obtained from studies in which in vivo metabolism and bioactivation of $^{14}CCl_4$ was examined in rats pretreated with $CoCl_2$.[18] The uptake, metabolism, and bioactivation of CCl_4 was not significantly altered in $CoCl_2$-treated rats known to have highly decreased liver microsomal cytochrome P-450 content.

Additional experimental evidence indicating that actual liver injury observed in rats receiving a high dose of CCl_4 was identical regardless of whether prior protective dose was administered led to a reexamination of the mechanism underlying CCl_4 autoprotection.[54] A systematic time-course study in which biochemical, histopathological parameters as well as animal survival were examined revealed a critical role for the augmented and sustained tissue repair in the liver.[54] The protective dose-stimulated tissue repair results in augmented and sustained hepatocellular regeneration and tissue repair enabling the autoprotected rats to overcome the same massive liver injury, which is ordinarily irreversible and leads to hepatic failure followed by animal death.[54-58] Recently a similar mechanism has been reported for thioacetamide autoprotection,[68] thioacetamide heteroprotection against acetaminophen,[69] and 2-butoxyethanol autoprotection.[42]

7. Selective Ablation of the Early-Phase Cell Division Response by Colchicine

Finally, the pivotal importance of the early-phase stimulation of hepatocellular division and tissue repair was tested with colchicine. With a carefully selected dose of colchicine, it was possible to selectively ablate the early-phase stimulation of mitosis associated with the administration of a low dose of CCl_4.[56,57] One single administration of colchicine at 1 mg/kg results in ablation of mitotic activity, the effect lasting only up to 12 hr, such that the second phase of cell division at 48 hr after the administration of CCl_4 is unperturbed.[55] At this dose colchicine does not cause any detectable liver injury, nor does it cause any adverse perturbation of hepatobiliary function.[56] Therefore, use of colchicine permits a very important experimental paradigm in which the early-phase cell division in response to a low dose of CCl_4 can be selectively ablated. The selective ablation of the early-phase cell division resulted in a prolongation of limited liver injury associated with a low dose of CCl_4.[55] Ordinarily, intraperitoneal administration of 100 μL CCl_4/kg results in very limited liver injury, which is overcome by stimulated cell division and tissue repair,[20,21,32,33,43,44,49] within 24 hr. The prolongation of this limited injury lasts only for an additional 24 hr (up to 48 hr after CCl_4 injection), at which time the unperturbed second phase of cell division permits complete recovery to occur within the next 24 hr (by 72 hr after CCl_4 injection). This increased and prolonged CCl_4 injury is not accompanied by enhanced bioactivation of CCl_4.[55,57] Indeed, actual liver injury assessed by morphometric analysis or hepatocellular necrosis and ballooned cells is not enhanced during the first 12 hr in colchicine-treated rats, further indicating that enhancement of the mechanisms responsible for infliction of injury was not involved.[55,57] These findings underscore the pivotal role of the early-phase stimulation of cell division in the final outcome of toxicity associated with a low dose of CCl_4.

Another experimental paradigm permits a further test of how critical the early-phase cell division response is in the final outcome of hepatic injury. In the above-described experiments, the preservation of the second phase of cell division permits complete recovery by 72 hr. Administration of a large dose of CCl_4 permits one to experimentally interfere with this second phase of cell division. In such an experiment, the animals should not survive because of continued progression of toxicity. In other words, selective ablation of the early-phase cell division response in an autoprotection protocol should result in a denial of autoprotection. Indeed, 100% survival observed in an experimental protocol (100 μL CCl$_4$/kg administered 24 hr prior to the injection of 2.5 mL CCl$_4$/kg) is summarily denied by colchicine antimitosis.[57] This observation also provides very convincing experimental evidence for the newly proposed mechanism for the autoprotection phenomenon.[54,57] The mechanism underlying the autoprotection phenomenon is the ability of the liver tissue to respond by augmentation of tissue repair through cell division induced by the protective dose.[54]

Role of Proto-Oncogene Regulation of Apoptosis in the Progressive Phase of Liver Injury

In the analogy of a brush fire burning out of control to a forest fire, the mechanism that facilitates the rapid spread of the fire is the drying up of the greens surrounding the small brush fire, due to the heat. The rapid acceleration of the progressive phase of liver injury observed after the combination of chlordecone and CCl$_4$ may be explained by either the under expression of proto-oncogene *bcl-2* or overexpression of *c-myc* or the suppressor gene, *p53*. Proto-oncogene product, *bcl-2* is known to block apoptosis, while *c-myc* and *p53* are known to stimulate it (Table 8.7). Stimulatory effects of *c-myc* and *p53* on apoptosis and the inhibitory effects of *bcl-2* have been observed in a variety of experimental conditions.[70-74] Thus, under conditions when the early-phase burst of cell division is permitted to occur, expression of *bcl-2* may result in stimulation of apoptosis (Figure 8.7). Such an event would augment prompt recovery from any limited injury, as seen after the administration of a low dose of CCl$_4$. Alternatively, insufficient expression of *bcl-2* would permit the continued stimulation of apoptosis as a result of *c-myc*, *p53* or other gene products known to stimulate apoptosis (Figure 8.7).

Early Gene Expression, Cell Priming, and Molecular Foundations of Hormesis

The sequence of intracellular steps from the onset of mitogenic stimulation to cell division is poorly understood. One critical event which is

Table 8.7. Genes in Control of Apoptosis

Gene	Cellular Location of Protein Product	Effect on Apoptosis
bcl-2	Mitochondrial membrane Nuclear envelope Endoplasmic reticulum	Blocks
myc	Nucleus	Stimulates
p53	Nucleus	Stimulates Mutant blocks

Adapted from Reference #35.

Mechanism of Recovery

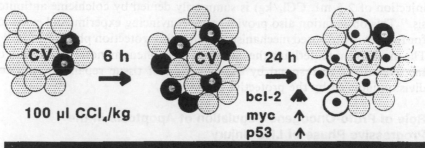

Mechanism of Progressive Injury

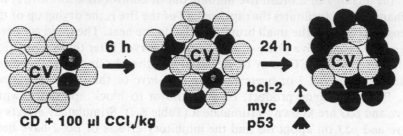

Figure 8.7.   Scheme illustrating the proposed mechanisms that permit a rapid recovery from limited injury of CCl₄ or accelerated progression of injury in animals treated with a combination of chlordecone and CCl₄. Proposed recovery mechanisms involve expression of proto-oncogene *bcl-2* to inhibit injury-stimulated apoptosis and the early-phase burst of cell division. In the absence of *bcl-2* expression, other gene products such as *cmyc* and *p53*, known to stimulate apoptosis would be expressed to contribute to the accelerated progression of injury. This, along with the lack of early-phase cell division, accounts for rapidly progressing liver injury.

thought to be essential in this process is the rapid and specific activation of immediate early genes, associated with competence for cell proliferation.[75] A prominent subset among these genes is the class of proto-oncogenes including *c-fos, c-myc,* and *c-jun* which encode for DNA-binding proteins that bind to DNA and transcriptionally activate additional genes needed for cell proliferation. One well studied in vivo model of cell proliferation has been liver regeneration after partial hepatectomy. Liver regeneration stimulated by partial hepatectomy provides an ideal in vivo model for the study of hepatocellular proliferation since hepatocytes are normally almost all arrested in the G_0 stage of the cell cycle (Figure 8.6). Surgical removal or toxic injury results in a prompt proliferative response that replaces the lost liver mass.[1,35,45,48,50,76,77] Cellular proliferation resulting from a 67% hepatectomy is associated with rapid rises in *c-fos,* and *c-myc* mRNA levels which peak at 30 min to 2 hr after surgery, and then quickly return to normal,[78] while *ras* gene expression peaks at the later time point of 24 to 48 hr.[79] These findings have led to the concept that early *c-fos* and *c-myc* expression primes the hepatocytes for subsequent proliferation.[76] Even though chemically-induced hepatotoxicity also results in hepatic regeneration, proto-oncogene expression associated with hepatotoxic models has received less attention. Work in this area has been confined to very few hepatotoxic models.[80-83] Working with CCl_4 as a hepatotoxic model, *c-myc* and *ras* expression,[80-82] and *c-fos, c-myc, c-jun, jun*-B, and *jun*-D[83] have been studied. Galactosamine also stimulates *c-fos, c-myc, c-jun, jun*-B, and *jun*-D expression, but at a slightly later time frame than observed for CCl_4.[83] The delayed stimulation of hepatocellular proliferation observed after galactosamine hepatotoxicity[84] may be a reflection of the mechanism of infliction of toxicity, rather than a difference in the timing of the proto-oncogene expression as a characteristic of the chemical. Another factor in the delayed cell proliferation response observed in the studies[83] might be the highly toxic doses employed, which are known to suppress the release of arrested G_2 cells, thereby ablating the early-phase cellular proliferation, as well as delay the second-phase cell proliferation response.[47,48,50] Although much investigative attention is needed in this area, it can be said that chemical injury results in activation of early proto-oncogene expression, a priming event required as an initial step. Primed cells are stimulated for S-phase synthesis and cell cycle progression is augmented by growth factors and other modulators of cellular proliferative and tissue repair activity.

A Two-Stage Model of Toxicity

An intriguing outcome of the work on the interactive toxicity of chlordecone + CCl_4 is the emergence of a concept, which permits the separation of the early events responsible for infliction of injury, from subsequent events which determine the final outcome of that injury (Figure 8.8). Hormetic mechanisms[85] are activated upon exposure to low levels of halomethanes.[9,20,21,32,33,43,66,69] Although the mechanisms responsible for triggering a dramatic mobilization of biochemical events leading to cellular proliferation within 6 hr after exposure to a subtoxic dose of CCl_4[9,22,32,33,43] are not understood, it is clear that these early events are the critical determinants of the final outcome of injury.[1,35,45,48] When this early phase of hepatocellular division is suppressed, as has been observed in animals pretreated with chlordecone,[20,32,33,43] a permissive and unabated progression of liver injury leading to massive coagulative hepatic necrosis is observed.[1,35,45,48] Likewise, experimentally, it has been demonstrated that restoring the tissue hormesis (Figure 8.9) inhibits progression of liver injury, permitting the tissue to overcome injury.

The central role of hormetic mechanisms in the final outcome of tissue injury becomes self-evident from the following lines of experimental evidence. Prior exposure to 225 ppm phenobarbital results in the potentiation of liver injury by the same subtoxic dose of CCl_4 employed in the chlordecone + CCl_4 interaction.[1,2,4,33] The quantitative measures of liver injury at

A Two - Stage Model of Toxicity

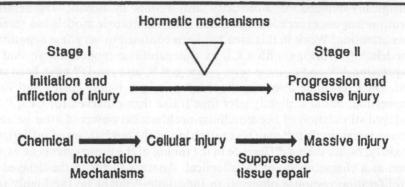

Figure 8.8.   Scheme illustrating the proposed two-stage model of toxicity. Stage I involves
infliction of cellular and/or tissue injury by intoxication mechanisms, which are
understood for many chemical and physical agents. When injury is inflicted by
a low dose of the offending agent (Stage I), hormetic mechanisms are
stimulated (such as cellular regeneration and tissue repair targeted for
restoration of tissue structure) and complete recovery from injury follows with
no additional toxic consequence. If hormetic mechanisms are suppressed or
ablated, the limited injury associated with exposure to a low dose of the
offending toxic agent would continue unabated resulting in progressive injury.
High doses of toxic agents can cause ablation of the hormetic mechanism, as
in the case of high dose of CCl_4, which results in ablation of the early-phase
hormetic response. Another example is the ablation of the early-phase
hormesis exemplified by the interactive toxicity of chlordecone and the
halomethane solvents. Adapted from Reference #50.

24 hr after the administration of CCl_4 indicate that the tissue injury is either
equivalent to or slightly greater than that seen in chlordecone + CCl_4
interaction.[2] Left alone, the animals undergoing the toxicity of phenobarbital + CCl_4 combination recover, while those experiencing the chlordecone
+ CCl_4 combination do not.[1,4,35,45,48] While the enhanced liver injury
observed with the toxicity of phenobarbital + CCl_4 is consistent with the
increased bioactivation of CCl_4,[2,14] recovery from this injury is consistent
with the unabated hepatocellular proliferation and tissue repair.[33,43] Delayed
hepatocellular regeneration and tissue repair from the normal 6 hr to 24–36
hr[1,33] is the only consequence on Stage II of CCl_4 toxicity. Nevertheless, the
highly stimulated early phase of tissue repair at 24 hr enables the restoration
of hepatolobular structure and function,[1,35,45,48,50] and thereby animal survival. These observations provide additional support for the concept of two
distinct stages of chemical toxicity (Figure 8.9).

Induction of liver regeneration 36 to 48 hr after the administration of a
toxic dose of CCl_4 is well established.[86-88] The existence of an early phase of
cell division (6 hr) was revealed only through experiments with low, subtoxic dose of CCl_4.[20,21,32,33] In fact, administration of a large, toxic dose of
CCl_4 (2.5 mL/kg) results in complete suppression of this early phase of cell
division,[44,51,54] indicating that the toxicity associated with a large dose is due

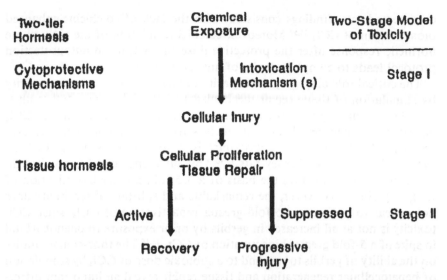

Two-tier
Hormesis
_____

Cytoprotective
Mechanisms

Chemical
Exposure

Intoxication
Mechanism (s)

Cellular Inury

Cellular Proliferation
Tissue Repair

Tissue hormesis

Active

Recovery

Suppressed

Progressive
Injury

Two-Stage Model
of Toxicity
_____

Stage I

Stage II

Figure 8.9.   Scheme illustrating the concept of separating those mechanisms that are responsible for the infliction of cellular and tissue injury from those that come to follow these events. Intoxication mechanisms result in infliction of injury during Stage I of toxicity. During Stage I of toxicity, tissue hormetic mechanisms are stimulated in an attempt to overcome injury. If these hormetic mechanisms are unperturbed, recovery occurs. Interference with these mechanisms results in uncontrollable progression of injury much like an unquenched brushfire, resulting in Stage II of toxicity. Adapted from Reference #50.

to the abolishment of this critical early phase stimulation of tissue repair.[1,35,45,48] Therefore, it is possible to ablate the early phase of hepatocellular regeneration and tissue repair ordinarily stimulated by a low dose of CCl_4, making it in essence a toxic dose. Administration of the same dose to animals prestimulated by partial hepatectomy so that they have the ongoing hepatocellular proliferation and tissue repair, results in a remarkable and substantial protection from liver injury and lethality.[51] Likewise, administration of a large lethal dose of CCl_4 to animals receiving a smaller dose to stimulate cell division and tissue repair results in complete protection.[54,57] Such protection is not due to decreased bioactivation of CCl_4.[18,55]

The importance of the stimulation of tissue repair as an event independent of Stage I of chemical toxicity can be illustrated by other elegant experimental approaches. Experimental interference with the early phase of hepatocellular proliferation leads to prolonged and enhanced liver injury of an ordinarily subtoxic dose of CCl_4. Studies with colchicine antimitosis,[55,57] wherein colchicine dose administered selectively ablates the early phase of hepatocellular division (6 hr) without interfering with the second phase of hepatocellular regeneration (48 hr), have shown a prolongation of liver injury. Neither liver injury measured through serum enzyme elevations nor by morphometric analysis of necrosis was increased at 6 or 12 hr in colchicine treated rats, findings consistent with the lack of colchicine-enhanced bioactivation of CCl_4.[55,57] Moreover, colchicine ablation of the early-phase hormetic response after the protective dose of CCl_4 in an autoprotection protocol leads to complete denial of autoprotection.

The critical role played by the capacity to respond to CCl_4-hepatotoxicity by stimulation of tissue repair mechanisms at an early time point is illustrated by examining species and strain differences in susceptibility to CCl_4 injury. Mongolian gerbils are extremely sensitive to halomethane hepatotoxicity.[15,16,53,89] Gerbils are approximately 35-fold more sensitive to CCl_4 toxicity than Sprague-Dawley rats.[15,16] This difference in CCl_4 toxicity can be seemingly explained on the basis of a 3.5-fold greater bioactivation of CCl_4 in gerbils.[15] However, the remarkable and substantial sensitivity does not appear to be due to 3.5-fold greater bioactivation of CCl_4, since CCl_4 toxicity is not at all increased in gerbils by prior exposure to phenobarbital in spite of a 5-fold greater bioactivation of CCl_4.[15,16] The time-course studies on the ability of gerbils to respond to a subtoxic dose of CCl_4 by stimulation of hepatocellular regeneration and tissue repair reveal an important difference in the biology of the hormetic mechanisms between gerbils and rats.[16] The early-phase stimulation of tissue repair in the liver does not manifest itself in gerbils and the second phase occurs approximately 40 hr after the administration of CCl_4.[16,53] In the absence of the biological mechanism to arrest the progression of liver injury (Figure 8.9), the liver injury might be expected to permissively progress much like an unquenched brushfire.

Evidence in support of the concept that species differences in chemical toxicity might depend on the differences in the promptness in initiating tissue repair mechanisms among various species comes from another aspect of the interactive toxicity of chlordecone + CCl_4. While gerbils are extremely sensitive to CCl_4, this sensitivity cannot be further increased by prior exposure to chlordecone.[15,16,53,89] Since substantial evidence supports the concept that suppression of the early phase of hepatocellular regeneration and tissue repair is the mechanism for the permissive progression of liver injury in the chlordecone + CCl_4 interaction,[1,35,45,48,49] lack of this early phase response in the gerbil would be consistent with extremely high sensitivity of gerbils to CCl_4 on the one hand, and a lack of potentiation of CCl_4 toxicity by prior exposure to chlordecone on the other.[15,16] This concept has received additional support recently through partial hepatectomy experiments.[53]

The toxicity of chlordecone + $CHCl_3$ combination has been demonstrated in murine species.[6-9] Stimulation of hepatocellular regeneration and tissue repair after a subtoxic dose of $CHCl_3$ allows the mice to overcome the liver injury associated with that dose of $CHCl_3$.[9] By lowering the dose of $CHCl_3$ used in the chlordecone + $CHCl_3$ studies,[8] it is possible to demonstrate potentiation of liver injury, but without the lethality.[9] Such an experimental protocol vividly reveals a decisive role played by the stimulated tissue repair mechanisms in overcoming liver injury[9] and the separation of these mechanisms (Stage II) from the inflictive phase (Stage I) of chemical injury (Figure 8.9).

The importance of stimulated tissue repair mechanisms in overcoming liver injury has also been demonstrated through examination of the mechanistic basis for significant strain differences in mice.[90,91] A SJL/J strain of mice, known to be least susceptible to CCl_4 toxicity was shown to possess more prompt and efficient tissue repair mechanisms, which permit augmented recovery, while the BALB/c strain known to be more susceptible was shown to possess less efficient tissue repair mechanisms, resulting in retarded recovery.[90] The F_1 cross between these two strains was shown to be intermediate in susceptibility.[91] A careful histopathological evaluation revealed that while the time course of the appearance in injury was quite similar (Stage I, Figure 8.8), significant differences in tissue repair mechanisms between these strains could account for the strain differences in CCl_4 toxicity.[90,91] While the time course of the inflictive phase of injury in the F_1 (SJL/J × BALB/c) was similar to the two parent strains, the tissue repair was at the intermediate level of augmented (SJL/J) and retarded (BALB/c) recovery.

With the advent of the finding that a low dose of CCl_4 is not toxic, not so much because it does not initiate tissue injury, but because of the stimulated tissue repair mechanisms,[49] it became apparent that the stimulation of the early phase of hepatocellular regeneration is in essence an endogenous hormetic mechanism, recruited to overcome tissue injury. One implication of this finding is its possible role in the phenomenon of CCl_4 autoprotection.[54-57] Circumstantial evidence, wherein hepatic microsomal cytochrome P-450 was decreased by $CoCl_2$ administration to 40% of the normal level did not result in decreased CCl_4 liver injury,[49] suggested the possibility that mechanism(s) other than decreased cytochrome P-450 might be involved in CCl_4 autoprotection. Recent studies reveal a critical role for the hepatocellular regeneration and tissue repair stimulated by the low protective dose administration.[54] Essentially, the protective dose serves to stimulate tissue repair mechanisms[18,20,21,32,43] so that even before the large dose known to abolish the early phase stimulation of tissue repair[44] is administered, the tissue repair mechanisms are already in place, resulting in augmentation of tissue repair sufficient to tip the balance between injury and recovery in favor of the latter.[54] This experimental model represents another example wherein a selective augmentation of the tissue hormetic mechanism (Stage II, Figure 8.8) independent of the inflictive phase of toxicity (Stage I, Figure 8.8), can dramatically alter the ultimate outcome of toxic injury (Figure 8.9).

Another line of evidence to implicate the importance of the hormetic mechanisms in determining the final outcome of chemical toxicity comes from experiments designed to understand the mechanisms responsible for the failure of the tissue regenerative and repair mechanism in the interactive toxicity of chlordecone + CCl_4. Much evidence is available to implicate insufficient availability of cellular energy at a time when cell division should have taken place.[20,21,30-33] A remarkable and irreversibly precipitous decline in glycogen levels in the liver,[21,26] a rise in hepatocellular Ca^{2+},[22-25] a consequent stimulation of phosphorylase a activity, leading to an equally precipitous decline in hepatic ATP,[26,28] are events consistent with the failure of hepatocellular regeneration in the chlordecone + CCl_4 interaction. Only marginal and transient decline in ATP levels in the interactive hepatotoxicity of phenobarbital + CCl_4 and mirex + CCl_4[28] are consistent with only a postponement of hepatocellular regeneration leading to transiently increased liver injury followed by complete recovery.[33] The concept of insufficient hepatocellular energy being linked to failure of hepatocellular regeneration and tissue repair has gained support from experiments in which the administration of external source of energy resulted in augmented ATP levels and significant protection.[29,30,51] Catechin (cyanidanol), known to increase hepatic ATP levels, protects against the lethal effect of chlordecone + CCl_4.[29,30] Protection by catechin is accompanied by a restored stimulation of hepatolobular repair and tissue healing.[30] The most interesting aspect of catechin protection against the interactive toxicity of chlordecone + CCl_4 is that protection does not appear to be the result of decreased infliction of hepatic injury,[29,30] as evidenced by a lack of difference in injury up to 24 hr after CCl_4 administration.[30] These observations provide substantial evidence for the separation of Stage I of toxicity responsible for the infliction of tissue injury from the Stage II events responsible for the final outcome of tissue injury.[48]

Abundant opportunities to test the two-stages of model toxicity are available. Many chemicals have been reported to induce hepatocellular regeneration at relatively modest doses, some of which are listed in Table 8.8. Opportunities to test the conceptual framework being put forth here are available through additional investigations with these models of tissue injury as well as scores of other models in other tissues and organs.

Table 8.8. Chemicals Reported to Stimulate Hepatocellular Proliferation

Chemicals	References
1. Acetaminophen	92
2. Allyl alcohol	92, 93
3. a-Naphthyl isothiocyanate	94, 95
4. Bromotrichloromethane	96–98
5. Carbon tetrachloride	1, 2, 20, 21, 49
6. Chloroform	8, 9
7. Ethylene dibromide	99
8. Galactosamine	84, 100, 101
9. Thioacetamide	102, 103

Adapted from Reference #35.

Modulators of the Two Stages of Toxicity

A number of modulators of toxicity are known to increase or decrease Stage I of toxicity (Table 8.9). Likewise, a number of separate modifiers of Stage II of toxicity are known (Table 8.10). A variety of experimental evidence can be cited for the existence of Stage II of toxicity (Table 8.11). These are dissociable from the Stage I events that result in the initiation of the infliction of injury.

IMPLICATIONS FOR ASSESSMENT OF RISK TO PUBLIC HEALTH

Establishing that the initial toxic or injurious events, regardless of how they are caused, can be separated from the subsequent events that determine the ultimate outcome of injury, offers promising opportunities for developing new avenues for therapeutic intervention, with the aim of restoring the hormetic tissue repair mechanisms. Such a development will open up avenues for two types of measures to protect public health. The presently used principle is to decrease the injury by interfering with Stage I of toxicity by treatment with an antidote, which either prevents further injury or

Table 8.9. Stage I of Chemical Injury

■ Bioactivation-dependent toxicity	■ Direct toxicity
– Increased by inducers	– Modulators of disposition alter
– Decreased by inhibitors	toxicity
– Balance between intoxication and	
detoxication	
– Modulators of chemical disposition	

Adapted from Reference #35.

Table 8.10. Stage II of Chemical Toxicity

■ Suppressed tissue repair:
  – Amplifiers of toxicity
  – Larger dose of a chemical
  – Depletion of cellular energy

Adapted from Reference #35.

Table 8.11. Evidence for Existence of Stage II

■ Toxic chemicals stimulate cell division.
■ Growing livers are resilient to chemical toxicity.
■ Stimulation of cell division affords protection.
■ Newly divided cells are resistant to toxicity.
■ Species differences.
■ Increased bioactivation increases infliction of injury, but recovery occurs.
■ Protective agents stimulate cell division.
■ Autoprotection is due to stimulated cell division.

Adapted from Reference #35.

decreases already inflicted injury. The second is wherein tissue repair and healing mechanisms could be enhanced not only to obtund the progression of injury, but also to simultaneously augment recovery from that injury.

In addition to these opportunities, the two-stage concept of chemical toxicity also embodies implications of significant interest in the assessment of risk from exposure to toxic chemicals. The existence of a threshold for chemical toxicity is evident, as indicated by the stimulation of tissue repair mechanism directed to tissue healing and recovery observed after the administration of subtoxic levels of toxic chemicals, when exposure involves singular chemicals. The existence of a two-level or two-stage threshold is apparent from the two-tier hormetic response: one threshold for each stage of the two-stage model. Generally speaking the threshold for Stage I of toxicity must lie in the cytoprotective mechanisms (cellular hormesis). The threshold for Stage II of toxicity appears to be in the tissue's ability to respond promptly by augmenting tissue healing mechanisms. These thresholds may be quantitatively the same or different.

From a public health perspective, exposure to singular chemicals is seldom involved. Multiple exposures to chemical combinations and/or singular components simultaneously, intermittently, or sequentially are almost always the rule. In this regard, antagonistic interactive toxicity or inconsequential interactions are also of interest. Of greater interest from a public health perspective is the finding that the hormetic mechanisms which constitute the threshold for physical or chemical toxicity can be mitigated by other chemical and physical agents, resulting in highly accentuated toxicity.

Of significantly greater interest is the need to take into account the hormetic mechanisms operating particularly at low levels of exposure to chemicals, in the assessment of risk from exposures to combinations of chemicals at low doses. The recognition of the existence of cellular and tissue hormesis provides a mechanistic basis to recognize thresholds for toxic effects, thereby permitting us to take into consideration the lack of recognizable adverse health effects at low levels of exposure to chemicals in our environment.

ACKNOWLEDGMENTS

The author's research received grant support from the Department of the Air Force AFOSR-88-0009, the Harry G. Armstrong Aerospace Medical Research Laboratory through USEPA CR-814053, by the 1988 Burroughs Wellcome Toxicology Scholar Award, and by The Burroughs Wellcome Fund.

REFERENCES

1. Mehendale, H.M. "Potentiation of Halomethane Hepatotoxicity by Chlordecone: A Hypothesis for the Mechanism," *Med. Hypoth.* 33:289–299 (1990).
2. Mehendale, H.M. "Potentiation of Halomethane Hepatotoxicity: Chlordecone and Carbon Tetrachloride," *Fundam. Appl. Toxicol.* 4:295–308 (1984).

3. Curtis, L.R., W.L. Williams and H.M. Mehendale. "Potentiation of the Hepatotoxicity of Carbon Tetrachloride Following Pre-Exposure to Chlordecone (Kepone®) in the Male Rat," *Toxicol. Appl. Pharmacol.* 51:283–293 (1979).
4. Klingensmith, J.S., and H.M. Mehendale. "Potentiation of CCl₄ Lethality by Chlordecone," *Toxicol. Lett.* 11:149–154 (1982).
5. Agarwal, A.K., and H.M. Mehendale. "Potentiation of CCl₄ Hepatotoxicity and Lethality by Chlordecone in Female Rats," *Toxicology* 26:231–242 (1983).
6. Hewitt, W.R., H. Miyajima, M.G. Cote, and G.L. Plaa. "Acute Alteration of Chloroform-Induced Hepato- and Nephrotoxicity by Mirex and Kepone®," *Toxicol. Appl. Pharmacol.* 48:509–517 (1979).
7. Hewitt, L.A., C. Palmason, S. Masson and G.L. Plaa. "Evidence for the Involvement of Organelles in the Mechanism of Ketone®-Potentiated Chloroform-Induced Hepatotoxicity," *Liver* 10:35–48 (1990).
8. Purushotham, K.R., V.G. Lockard, and H.M. Mehendale. "Amplification of Chloroform Hepatotoxicity and Lethality by Dietary Chlordecone in Mice," *Toxicol. Pathol.* 16:27–34 (1988).
9. Mehendale, H.M., K.R. Purushotham, and V.G. Lockard. "The Time-Course of Liver Injury and 3H-Thymidine Incorporation in Chlordecone-Potentiated CHCl₃ Hepatotoxicity," *Exp. Mol. Pathol.* 51:31–47 (1989).
10. Klingensmith, J.S., and H.M. Mehendale. "Potentiation of Brominated Halomethane Hepatotoxicity by Chlordecone in the Male Rat," *Toxicol. Appl. Pharmacol.* 61:429–440 (1981).
11. Agarwal, A.K., and H.M. Mehendale. "Potentiation of Bromotrichloromethane Hepatotoxicity and Lethality by Chlordecone Pre-Exposure in the Rat," *Fundam. Appl. Toxicol.* 2:161–167 (1982).
12. Mehendale, H.M., and V.G. Lockard. "Effect of Chlordecone on the Hepatotoxicity of 1,1,2-Trichloroethylene and Bromobenzene," *The Toxicologist* 2:37 (1982).
13. Fouse, B.L., and E. Hodgson. "Effect of Chlordecone and Mirex on the Acute Hepatotoxicity of Acetaminophen in Mice," *Gen. Pharmacol.* 18:623–630 (1987).
14. Mehendale, H.M., and J.S. Klingensmith. "In Vivo Metabolism of CCl₄ by Rats Pretreated with Chlordecone, Mirex or Phenobarbital," *Toxicol. Appl. Pharmacol.* 93:247–256 (1988).
15. Cai, Z., and H.M. Mehendale. "Lethal Effects of CCl₄ and Its Metabolism by Gerbils Pretreated with Chlordecone, Phenobarbital and Mirex," *Toxicol. Appl. Pharmacol.* 104:511–520 (1990).
16. Cai, Z., and H.M. Mehendale. "Hepatotoxicity and Lethality of Halomethanes in Mongolian Gerbils Pretreated with Chlordecone, Phenobarbital or Mirex," *Arch. Toxicol.* 65:204–212 (1991).
17. Harris, R.N., and M.W. Anders. "2-Propanol Treatment Induces Selectively the Metabolism of Carbon Tetrachloride to Phosgene: Implications for Carbon Tetrachloride Hepatotoxicity," *Drug Metab. Dispos.* 9:551–556 (1981).
18. Young, R.A., and H.M. Mehendale. "Carbon Tetrachloride Metabolism in Partially Hepatectomized and Sham Operated Rats Pre-Exposed to Chlordecone," *J. Biochem. Toxicol.* 4:211–219 (1989).
19. Davis, M.E., and H.M. Mehendale. "Functional and Biochemical Correlates of Chlordecone Exposure and Its Enhancement of CCl₄ Hepatotoxicity," *Toxicology* 15:91–103 (1980).

20. Lockard, V.G., H.M. Mehendale and R.M. O'Neal. "Chlordecone-Induced Potentiation of Carbon Tetrachloride Hepatotoxicity: A Light and Electron Microscopic Study," *Exp. Mol. Pathol.* 39:230–245 (1983).

21. Lockard, V.G., H.M. Mehendale, and R.M. O'Neal. "Chlordecone-Induced Potentiation of Carbon Tetrachloride Hepatotoxicity: A Morphometric and Biochemical Study," *Exp. Mol. Pathol.* 39:246–256 (1983).

22. Agarwal, A.K., and H.M. Mehendale. "CCl_4-Induced Alterations in Ca^{2+} Homeostasis in Chlordecone and Phenobarbital Pretreated Animals," *Life Sci.* 34:141–148 (1984).

23. Agarwal, A.K., and H.M. Mehendale. "Excessive Hepatic Accumulation of Intracellular Ca^{2+} in Chlordecone Potentiated CCl_4 Toxicity," *Toxicology* 30:17–24 (1984).

24. Carmines, E.L., R.A. Carchman, and J.F. Borzelleca. "Kepone®: Cellular Sites of Action," *Toxicol. Appl. Pharmacol.* 49:543–550 (1979).

25. Agarwal, A.K., and H.M. Mehendale. "Effect of Chlordecone on Carbon Tetrachloride-Induced Increase in Calcium Uptake in Isolated Perfused Rat Liver," *Toxicol. Appl. Pharmacol.* 83:342–348 (1986).

26. Kodavanti, P.R.S., U.P. Kodavanti, and H.M. Mehendale. "CCl_4-Induced Alterations in Hepatic Calmodulin and Free Calcium Levels in Rats Pretreated with Chlordecone," *Hepatology* 13:230–238 (1991).

27. Kodavanti, P.R.S., V.C. Rao, and H.M. Mehendale. "Loss of Calcium Homeostasis Leads to Progressive Phase of Chlordecone-Potentiated Carbon Tetrachloride Hepatotoxicity," *Toxicol. Appl. Pharmacol.* 122:77–87 (1993).

28. Kodavanti, P.R.S., U.P. Kodavanti, and H.M. Mehendale. "Altered Hepatic Energy Status in Chlordecone (Kepone®) Potentiation of CCl_4 Hepatotoxicity," *Biochem. Pharmacol.* 40:859–866 (1990).

29. Soni, M.G., and H.M. Mehendale. "Protection from Chlordecone-Amplified Carbon Tetrachloride Toxicity by Cyanidanol: Biochemical and Histological Studies," *Toxicol. Appl. Pharmacol.* 108:46–57 (1991).

30. Soni, M.G., and H.M. Mehendale. "Protection from Chlordecone-Amplified Carbon Tetrachloride Toxicity by Cyanidanol: Regeneration Studies," *Toxicol. Appl. Pharmacol.* 108:58–66 (1991).

31. Soni, M.G., and H.M. Mehendale. "Adenosine Triphosphate Protection of Chlordecone-Amplified CCl_4 Hepatotoxicity and Lethality," *J. Hepatol.* 20:267–274 (1994).

32. Kodavanti, P.R.S., U.M. Joshi, R.A. Young, A.N. Bell, and H.M. Mehendale. "Role of Hepatocellular in Chlordecone-Potentiated Hepatotoxicity of Carbon Tetrachloride," *Arch. Toxicol.* 63:367–375 (1989).

33. Kodavanti, P.R.S., U.P. Kodavanti, O.M. Faroon, and H.M. Mehendale. "Correlation of Hepatocellular Regeneration and CCl_4-Induced Hepatotoxicity in Chlordecone, Mirex or Phenobarbital Pretreated Rats," *Toxicol. Pathol.* 20:556–569 (1992).

34. Slater, R.F. "Free Radicals and Tissue Injury: Fact and Fiction," *Br. J. Cancer (Suppl)* 8:5–10 (1987).

35. Mehendale, H.M. "Mechanism of the Interactive Amplification of Halomethane Hepatotoxicity and Lethality by Other Chemicals," in *Toxicology of Chemical Mixtures: From Mechanisms to Real Life Examples*, Chapter 13, R.S.H. Yang, Ed., (Chelsea, MI: Lewis Publishers, 1994) in press.

36. Ruch, R.J., J.E. Klaunig, N.E. Schultz, A.B. Askari, D.A. Lacher, M.A.

Pereira, and P.J. Goldblatt. "Mechanisms of Chloroform and Carbon Tetrachloride Toxicity in Primary Cultured Mouse Hepatocytes," *Environ. Health Perspect.* 69:301–305 (1986).

37. Roberts, E., B.M. Ahluwalia, G. Lee, C. Chan, D.S.R. Sarma, and E. Farber. "Resistance to Hepatotoxins Acquired by Hepatocytes During Liver Regeneration," *Cancer Res.* 43:28–34 (1983).

38. Chang, L.W., M.A. Pereira, and J.E. Klaunig. "Cytotoxicity of Halogenated Alkanes in Primary Cultures of Rat Hepatocytes from Normal Partial Hepatectomized and Preneoplastic/Neoplastic Liver," *Toxicol. Appl. Pharmacol.* 80:274–280 (1985).

39. Ruch, R.J., J.E. Klaunig, and M.A. Pereira. "Selective Resistance to Cytotoxic Agents in Hepatocytes Isolated from Partially Hepatectomized and Neoplastic Mouse Liver," *Cancer Lett.* 26:295 (1985).

40. Cai, C., and H.M. Mehendale. "Role of Ongoing Versus Stimulated Hepatocellular Regeneration in Resiliency to Amplification of CCl_4 Toxicity by Chlordecone," *FASEB J.* 5:A1248 (1991).

41. Cai, Z., and H.M. Mehendale. "Resiliency to Amplification of Carbon Tetrachloride Hepatotoxicity by Chlordecone During Postnatal Development in Rats," *Pediatr. Res.* 33:225–232 (1993).

42. Digavalli, S., and H.M. Mehendale. "Autoprotection Against Lethality of 2-Butoxyethanol," *FASEB J.* 7:A803 (1993).

43. Kodavanti, P.R.S., U.M. Joshi, V.G. Lockard, and H.M. Mehendale. "Chlordecone (Kepone®)-Potentiated Carbon Tetrachloride Hepatotoxicity in Partially Hepatectomized Rats. A Histomorphometric Study," *J. Appl. Toxicol.* 9:367–375 (1989).

44. Kodavanti, P.R.S., U.M. Joshi, R.A. Young, E.F. Meydrech, and H.M. Mehendale. "Protection of Hepatotoxic and Lethal Effects of CCl_4 by Partial Hepatectomy," *Toxicol. Pathol.* 17:494–506 (1989).

45. Mehendale, H.M. "Amplification of Hepatotoxicity and Lethality of CCl_4 and $CHCl_3$ by Chlordecone," *Rev. Biochem. Toxicol.* 10:91–138 (1989).

46. Soni, M.G., and H.M. Mehendale. "Hepatic Failure Leads to Lethality of Chlordecone-Amplified Hepatotoxicity of Carbon Tetrachloride," *Fundam. Appl. Toxicol.* 22:442–450 (1993).

47. Calabrese, E.J., L.A. Baldwin, and H.M. Mehendale. "G_2 Subpopulation in Rat Liver Induced into Mitosis by Low Level Exposure to Carbon Tetrachloride: An Adaptive Response," *Toxicol. Appl. Pharmacol.* 120:1–7 (1993).

48. Mehendale, H.M. "Role of Hepatocellular Regeneration and Hepatolobular Healing in the Final Outcome of Liver Injury: A Two-Stage Model of Toxicity," *Biochem. Pharmacol.* 42:1155–1162 (1991).

49. Bell, A.N., R.A. Young, V.G. Lockard, and H.M. Mehendale. "Protection of Chlordecone-Potentiated Carbon Tetrachloride Hepatotoxicity and Lethality by Partial Hepatectomy," *Arch. Toxicol.* 61:392–405 (1988).

50. Mehendale, H.M. "Biochemical Mechanisms of Biphasic Dose-Response Relationships: Role of Hormesis," in *Biological Effects of Low Level Exposures to Chemicals and Radiation*, Chapter 4, E.J. Calabrese, Ed., (Chelsea, MI: Lewis Publishers, 1992), pp. 59–94.

51. Rao, S.B., and H.M. Mehendale. "Protective Role of Fructose 1,6-Bisphosphate During CCl_4 Hepatotoxicity in Rats," *Biochem. J.* 262:721–725 (1989).

52. Mehendale, H.M., Z. Cai, and S.,D. Ray. "Paradoxical Toxicity of CCl_4 in Isolated Hepatocytes from Chlordecone, Phenobarbital and Mirex Pretreated Rats," *In Vitro Toxicol.* 4:187–196 (1991).

53. Cai, Z., and H.M. Mehendale. "Prestimulation of Hepatocellular Regeneration by Partial Hepatectomy Decreases Toxicity of CCl_4 in Gerbils," *Biochem. Pharmacol.* 42:633–644 (1991).

54. Thakore, K.N., and H.M. Mehendale. "Role of Hepatocellular Regeneration in Carbon Tetrachloride Autoprotection," *Toxicol. Pathol.* 19:47–58 (1991).

55. Rao, C.V., and H.M. Mehendale. "Prolongation of Carbon Tetrachloride Toxicity by Colchicine Antimitosis," *Arch. Toxicol.* 67:392–400 (1993).

56. Rao, C.V., and H.M. Mehendale. "Effect of Colchicine on Hepatobiliary Function in CCl_4 Treated Rats," *Biochem. Pharmacol.* 42:2223–2232 (1991).

57. Rao, C.V., and H.M. Mehendale. "Colchicine Antimitosis Abolishes CCl_4 Autoprotection," *Toxicol. Pathol.* 19:597–606 (1991).

58. Mehendale, H.M., K.N. Thakore, and V.C. Rao. "Autoprotection: Stimulated Tissue Repair Permits Recover from Injury," *J. Biochem. Toxicol.* 9: in press (1994).

59. Glende, Jr., E.A. "Carbon Tetrachloride-Induced Protection Against Carbon Tetrachloride Toxicity: The Role of the Liver Microsomal Drug-Metabolizing System," *Biochem. Pharmacol.* 21:1697–1702 (1972).

60. Dambrauskas, T., and H.H. Cornish. "Effect of Pretreatment of Rats with Carbon Tetyrachloride in Tolerance Development," *Toxicol. Appl. Pharmacol.* 17:3–97 (1970).

61. Ugazio, G., R.R. Koch, and R.O. Recknagel. "Mechanism of Protection Against Carbon Tetrachloride by Prior Carbon Tetrachloride Administration," *Exp. Mol. Pathol.* 16:281–285 (1973).

62. Gerhard, H.J., B. Schultz, and W. Maurer. "Wirkung einor zweiten CCl_4-intoxikation aufdie CCl_4-geschadigte Leber de Mans," *Virchows Abk. B. Zellerpath.* 10:184–199 (1972).

63. Pound, A.W., and T.A. Lawson. "Reduction of Carbon Tetrachloride Toxicity by Prior Administration of a Single Small Dose in Mice and Rats," *Br. J. Exp. Pathol.* 56:172–179 (1975).

64. Recknagel, R.O., and E.A. Glende, Jr. "Lipid Peroxidation: A Specific Form of Cellular Injury," in *Handbook of Physiology*, Section 9, D.H.K. Lee, Ed., American Physiological Society, Bethesda, MD, (Baltimore, MD: Williams and Wilkins, 1977), pp. 591–601.

65. Sipes, I.G., G. Krishna, and J.R. Gillette. "Bioactivation of Carbon Tetrachloride, Chloroform and Bromotrichloromethane: Role of Cytochrome P-450." *Life Sci.* 20:1541–1548 (1974).

66. Klingensmith, S.J., and H.M. Mehendale. "Destruction of Hepatic Mixed Function Oxygenase Parameters by CCl_4 in Rats Following Acute Treatment with Chlordecone, Mirex and Phenobarbital," *Life Sci.* 33:2339–2348 (1984).

67. Klingensmith, S.J. "Metabolism of CCl_4 in Rats Pretreated with Chlordecone, Mirex and Phenobarbital," University of Mississippi Medical Center, PhD Dissertation, 1982.

68. Mangipudy, R.S., and H.M. Mehendale. "Thioacetamide Autoprotection," *FASEB J.* 7:A40 (1993).

69. Chanda, S., and H.M. Mehendale. "Heteroprotection by Thioacetamide Against Acetaminophen Lethality," *Toxicologist* 13:32 (1993).

70. Hockenbery, D.M., G. Nunez, C. Milliman, M. Alexander, R.D. Schreiber, and J.P. Korsmyer. "Bcl$_2$ is an Inner Mitochondrial Membrane Protein That Blocks Programmed Cell Death," *Nature* 348:334–337 (1990).

71. Hockenbery, D.M., M. Zutter, M. Hickey, M. Nahm, and S.J. Korsmeyer. "Bcl$_2$ Protein is Topographically Restricted in Tissues Characterized by Apoptotic Cell Death," *Proc Natl. Acad. Sci. (USA)* 18:6961–6965 (1991).

72. Williams, G.T. "Programmed Cell Death: Apoptosis and Oncogenesis," *Cell* 65:1097–1098 (1991).

73. Alnemri, E.S., T.F. Fernandes, S. Haldana, C.M. Croce, and G.L. Litwack. "Involvement of Bcl$_2$ in Glucocorticoid-Induced Apoptosis of Pre-B-Leukemias," *Cancer* 52:491–495 (1992).

74. Sinkovics, J.G. "Programmed Cell Death (Apoptosis): Its Virological and Immunological Connections (A Review)," *Acta Microbiol. Hung.* 38:321–324 (1991).

75. Rollins, B.J., and C.D. Stiles. "Regulation of *c-myc* and *c-fos* Proto-Oncogene Expression by Animal Cell Growth Factor," *In Vitro Cell Dev. Biol.* 24:81–84 (1988).

76. Fausto, N., and J.E. Mead. "Regulation of Liver Growth: Proto-Oncogenes and Transforming Growth Factors," *Lab Invest.* 48:224–230 (1989).

77. Michalopoulos, G.K. "Liver Regeneration: Molecular Mechanisms of Growth Control," *FASEB J.* 4:176–187 (1990).

78. Thompson, N.L., J.E. Mead, L. Braun, M. Goyette, P.R. Shank, and N. Fausto. "Sequential Proto-Oncogene Expression During Rat Liver Regeneration," *Cancer Res.* 46:3111–3117 (1986).

79. Makino, R., K. Hayashi, and T. Sugimura. "*c-myc* Transcript is Induced in Rat Liver at a Very Early Stage of Regeneration or by Cycloheximide Treatment," *Nature (Lond.)* 310:697–698 (1984).

80. Goyette, M., C.J. Petropolos, P.R. Shank, and N. Fausto. "Regulated Transcription of *cki-ras* and *c-myc* During Compensatory Growth of Rat Liver," *Mod. Cell Biol.* 4:1493–1498 (1984).

81. Herbst, H., S. Milani, D. Schuppan, and H. Stein. "Temporal and Spatial Patterns of Proto-Oncogene Expression at Early Stages of Toxic Liver Injury in the Rat," *Lab Invest.* 65:324–333 (1991).

82. Sasaki, Y., N. Hayashi, Y. Morita, T. Ito, A. Kasahara, H. Fusamoto, N. Sato, M. Tohyma, and T. Kamada. "Cellular Analysis of *c-Ha-ras* Gene Expression in Rat Liver After CCl$_4$ Administration," *Hepatology* 10:494–500 (1989).

83. Schmeidberg, P., L. Biempica, and M.J. Czaja. "Timing of Proto-Oncogene Expression in Rat Liver After CCl$_4$ Administration," *J. Cell Physiol.* 154:294–300 (1993).

84. Abdul-Hussain, S.K., and H.M. Mehendale. "Ongoing Hepatocellular Regeneration and Resiliency Toward Galactosamine Hepatotoxicity," *Arch. Toxicol.* 66:729–742 (1992).

85. Sagan, L. "On Radiation, Paradigms, and Hormesis," *Science* 245:574, then continued on 621 (1989).

86. Leevy, C.M., R.M. Hollister, R. Schmid, R.A. MacDonald, and C.S. Davidson. "Liver Regeneration in Experimental CCl$_4$ Intoxication," *Proc. Soc. Exp. Biol. Med.* 102:672–675 (1959).

87. Smuckler, E.A., M. Koplitz, and S. Sell. "Alpha-Fetoprotein in Toxic Liver Injury," *Cancer Res.* 36:4558–4561 (1976).

88. Nakata, R., I. Tsukamoto, M. Miyoshi, and S. Kojo. "Liver Regeneration After CCl₄ Intoxication in the Rat," *Biochem. Pharmacol.* 34:586–588 (1985).

89. Ebel, R.E., and E.A. McGrath. "CCl₄-Hepatotoxicity in the Mongolian Gerbil: Influence of Monooxygenase Induction," *Toxicol. Lett.* 22:205–210 (1984).

90. Bhathal, P.S., N.R. Rose, I.R. Mackay, and S. Whittingham. "Strain Differences in Mice in Carbon Tetrachloride-Induced Liver Injury," *Br. J. Exp. Pathol.* 64:524–533 (1983).

91. Biesel, K.W., M.N. Ehrinpreis, P.S. Bhathal, I.R. Mackay, and N.R. Rose. "Genetics of Carbon Tetrachloride-Induced Liver Injury in Mice. II Multigenic Regulation," *Br. J. Exp. Pathol.* 65:125–131 (1984).

92. Zieve, L., W.R. Anderson, C. Lyftogt, and K. Draves. "Hepatic Regenerative Enzyme Activity After Pericentral and Periportal Lobular Toxic Injury" *Toxicol. Appl. Pharmacol.* 86:147–158 (1986).

93. Zieve, L., W.R. Anderson, and D. Lafontaine. "Hepatic Failure Toxins Depress Liver Regenerative Enzymes After Periportal Injury with Allyl Alcohol in the Rat," *J. Lab. Clin. Med.* 111:725–730 (1988).

94. McLean, M.R., and K.R. Rees. "Hyperplacia of Bile-Ducts Induced by Alpha-naphthyl-iso-thiocyanate: Experimental Biliary Cirrhosis Free from Obstruction," *J. Pathol. Bacteriol.* 76:175–188 (1958).

95. Ungar, H., E. Moran, M. Eisner, and M. Eliakim. "Rat Intrahepatic Biliary Tract Lesions from Alpha-naphthyl-iso-thiocyanate," *Arch. Pathol.* 73:427–435 (1962).

96. Faroon, O.M., and H.M. Mehendale. "Bromotrichloromethane Hepatotoxicity. Role of Hepatocellular Regeneration in Recovery. Biochemical and Histopathological Studies in Control and Chlordecone Pretreated Male Rats," *Toxicol. Pathol.* 18:667–677 (1990).

97. Faroon, O.M., R.W. Henry, M.G. Soni, and H.M. Mehendale. "Potentiation of BrCCl₃ Hepatotoxicity by Chlordecone: Biochemical and Ultrastructural Study," *Toxicol. Appl. Pharmacol.* 110:185–197 (1991).

98. Thakore, K.N., M.L. Gargas, M.E. Andersen, and H.M. Mehendale. "PB-PK Derived Metabolism Constants, Hepatotoxicity, and Lethality of BrCCl₃ in Rats Pretreated with Chlordecone, Phenobarbital and Mirex," *Toxicol. Appl. Pharmacol.* 109:514–528 (1991).

99. Natchtomi, E., and E. Farber. "Ethylene Dibromide as a Mitogen for Liver," *Lab. Invest.* 38:279–283 (1978).

100. Lesch, R., W. Reutter, D. Keppler, and K. Decker. "Liver Restitution After Galactosamine Hepatitis: Autoradiographic and Biochemical Studies in Rats," *Exp. Mol. Pathol.* 12:58–69 (1970).

101. Kuhlmann, W.D., and K. Wurster. "Correlation of Histology and Alpha-Fetoprotein Surgence in Rat Liver Regeneration After Experimental Injury by Galactosamine," *Virchows Arch. Histol.* 387:47–57 (1980).

102. Gupta, D.N. "Acute Changes in the Liver After Administration of Thioacetamide," *J. Pathol. Bacteriol.* 72:183–192 (1956).

103. Reddy, J.K., M. Chiga, and D. Svoboda. "Initiation of Division of Cycle of Rat Hepatocytes Following a Single Injection of Thioacetamide," *Lab. Invest.* 20:405–411 (1969).

CHAPTER 9

Modulation of Toxicity By Diet: Implications for Response at Low Level Exposures

Angelo Turturro and Ronald Hart, Division of Biometry and Risk Assessment, National Center for Toxicological Research, U.S. Food and Drug Administration, Jefferson, Arkansas

INTRODUCTION

The model for the toxic effect of agents such as radiation or methyl bromide is almost always considered to be simple and direct, e.g., a damage is induced, usually in the genome, which leads to mutation, etc.[1] This approach may be appropriate for simple endpoints such as damage to the DNA (e.g., sister-chromatid exchanges[2]). However, complex endpoints, such as carcinogenesis and longevity, are the consequence of chains of events, usually over a long period, and almost always modifiable by numerous factors.[3] Actions related to compound exposure that modify these chains of events or the factors important to them will impact on the expression of exposure as an adverse health effect.

One of the most important factors in modifying chronic endpoints is diet. Diet, especially caloric restriction, is an effective inhibitor of spontaneous disease (especially cancer), induced toxicity,[4-8] and the only consistent enhancer of mammalian lifespan.[9,10] Because of the large magnitude of the organism's response to caloric restriction, any treatment which decreases caloric intake has the potential to decrease toxicity and extend lifespan. This potential treatment effect complicates the discussion of the monotonicity of the effects of low levels of radiation, both as a potential factor in evaluating the effects of an agent and as a paradigm which may shed some light on mechanisms important to nonmonotonic behavior.

HORMESIS

Following Sagan,[11] a broad definition of hormesis is used and will be considered to describe any observation encountered at low levels of exposure to agents which are not readily predictable from exposure to higher levels. In the context of the dose-response curve which describes the effect of the agent, hormesis can be described as nonmonotonic effects, which has the advantage of focusing attention on the single measure used in the dose-response curve. The measures commonly used include growth, cancer incidence, and longevity.[11] The following discussion will limit itself to longevity and carcinogenesis.

Hormesis and Longevity

In interpreting a change in mortality as a result of treatment, a question to consider is whether changes in the food consumption or utilization have occurred as a result of treatment. Using a chronic study (i.e., a two-year carcinogenesis study) on methyl bromide (MeBr) as an illustration example[12] (Figure 9.1), it can be seen that the effect of the lower exposure is to increase survival at 24 months, a nonmonotonic effect since the highest dose leads to increased mortality with exposure. This effect often occurs in toxicity studies, and has been suggested as evidence of hormesis,[1] although the doses of agents at which the effects occur cannot be considered low (often 1/4 the maximum tolerated dose). The effects of an agent on consumption are often not measured, and simply measuring consumption is complicated by food wastage and differential food utilization stimulated by treatment. However, body weight (BW) can be used as a biomarker for energy availability, i.e., a crude measure of food intake minus utilization

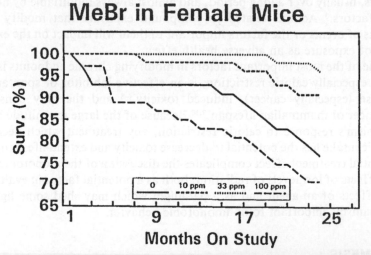

Figure 9.1. Survival in female mice at different exposures to methyl bromide (MeBr). Survival of female B6C3F1 mice exposed to different part-per-million (in air) exposures (5 times a week for eight hours) of methyl bromide (MeBr), with approximation per treatment. 0 is the control (no MeBr exposure). Placed on study at 5–6 weeks of age. Data plotted from Reference #12.

for ages young enough for the body weight not to be affected by concurrent pathology (for B6C3F1 mice less than 18 months of age). Evaluating BW (Figure 9.2), it can be seen that exposure to the lower levels of a toxicant in a test can lower body weight.

One of the most obvious and characteristic effects of caloric restriction is a reduction in body weight in restricted animals. Figure 9.3 is the effect on female B6C3F1 mice of a 40% reduction in caloric consumption. Note the body weight is lower throughout the lifespan. The effect of the restriction is to increase the survival at 111 weeks of age (roughly equivalent to the age of the animals when a chronic bioassay is terminated, since bioassays are started roughly 5–6 weeks after birth) from 80% for ad lib animals, to 95% in the restricted ones. The changes in body weight can be seen as a biomarker for effects similar to those seen with caloric restriction, suggesting that the hormetic effect may be a result of caloric restriction.

Using this information to interpret the results in MeBr-exposed mice, one possibility is that the increase in survival seen at the lower doses of agent (10 and 33 ppm) are a result of the effect on weight (this is also suggested by the larger effect at 33 ppm compared to 10 ppm, consistent with the effect on BW). This body weight loss may result from a variety of mechanisms, such as depressed appetite, changing metabolism, etc. At the 100 ppm dose, the toxic effects of the agent overcomes any positive effect generated by the large loss in BW seen, or the loss in body weight may represent the consequences of some illness induced by the toxicant. The competing mechanisms of lowered mortality resulting from weight loss being stimulated by the

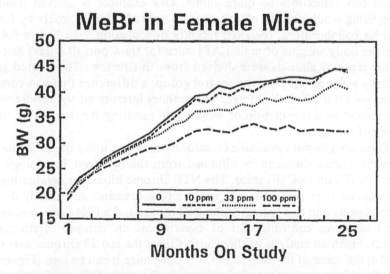

Figure 9.2. Body weight in female mice at different exposures to methyl bromide (MeBr). Average body weight of the same mice as in Figure 9.1. Data plotted from Reference #12.

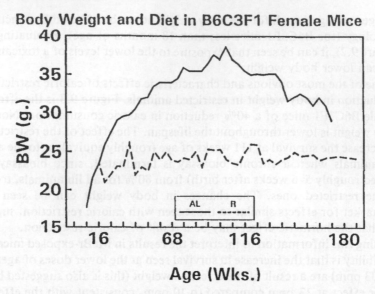

Figure 9.3. Body weight and diet in female mice. Body weight of female B6C3F1 mice eating either ad libitum or with a 40% caloric restriction. The restriction significantly decreases body weight throughout the lifespan, and the animal survival at 111 weeks of age significantly increases.

agent and increased mortality from the toxic effects of the agent can result in a nonmonotonic survival curve.

While the effect on caloric consumption is obvious in this study, the effect can sometimes be more subtle. One example is derived from the pioneering studies on the effects of low level radiation exposure by Lorenz and his collaborators, reviewed recently by Congdon.[13] In Figure 9.4, the average body weights of male LAF1 mice for the Cont. (0 r/day) and 0.11 r/day exposed animals were derived from this review. The treated group initially weighed less than the control group, a difference that was compensated for after 20 weeks of age. The authors interpreted the results of this experiment as a stimulation of weight gain resulting from exposure to low levels of radiation.

There are few data available on caloric restriction using the LAF1 mouse. However, some data can be obtained from the National Toxicology Program (NTP) on B6C3F1 mice. The NTP chronic bioassays, evaluating carcinogenesis after two years of exposure to a toxicant, are usually done in both genders and two species (a B6C3F1 mouse and a F-344 rat), and are the most extensive controlled set of experiments on chronic carcinogenesis extent. From an analysis of the controls from the last 13 chronic tests using feed as the route of toxicant in B6C3F1 male mice, it can be seen (Figure 9.5) that there is an inverse correlation between the percent alive at 24 months of age and the BW2, or the average body weight of the animals on the study at 2 months (8 weeks) following study initiation (approximately 14 weeks of

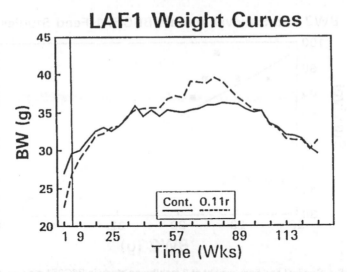

Figure 9.4. Body weight of LAF1 mice exposed to ionizing radiation. Data adapted from data presented in Reference #13 in mice exposed to a number of different radiation doses by Lorenz. Only data from the control dose (0 additional rad) and the lowest dose (0.11 rad) were graphed. The vertical line is the body weight at approximately 14 weeks of age (irradiation started at 2 months of age), equivalent to BW2, since animals in NTP studies are usually started at 5–6 weeks of age.

age). This observation is consistent with the many studies that have shown the extension of lifespan by caloric restriction,[10] which results in loss of BW when compared to controls when the restriction is instituted. Although, as noted in Figure 9.5, BW2 is a good estimator of survival at 24 months, BW12 (the average body weight at 12 months since study initiation) (Figure 9.6) is not. Using a linear regression derived from Figure 9.5,

$$S = -6.735 \times BW2(g) + 271.91 \qquad (9.1)$$

where S = % survival of a cohort at 24 months after study onset, and BW2 is as noted above. Using the body weights in the Lorenz study (denoted by arrows on Figure 9.5) of approximately 29.5 g for the control and 27 g for the 0.11 r/day treatment group, survivals of 90% for the treated group and 74% in the control can be calculated. Although the regression curve is derived from a different strain, this is consistent with the increase in mean survival time from 703 days (so the 24 month survival % is less than 50%) to 761 days (where the 24 month survival is greater than 50%) seen by Lorenz in the treated group. So, where the authors had considered the effects of irradiation as a stimulation of weight gain, reduced weight gain at a critical time in development (circa 4 months of age) could have been the mechanism for the hormesis effect associated with the low-level exposure to radiation.

Thus, caloric consumption and its pattern are major concerns in the

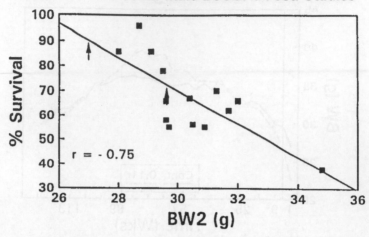

Figure 9.5.   Survival and body weight at 2 months on study in B6C3F1 male feed studies. Relationship of body weight at 2 months on study (app. 14 weeks of age) and survival at end of NTP feed studies. BW2 is the average body weight reported for controls at 8 weeks on study. Correlation coefficient is -0.75. Survival is percent survival at 105 weeks on study. Arrows point to body weights in LAF1 mouse study at 14 weeks of age. Data are derived from the latest available feed studies from the NTP Technical Reports (Carcinogenicity Bioassays) Series, i.e., TR419, TR412, TR407, TR387, TR365, TR353, TR341, TR337, TR324, TR320, TR315, TR309, and the NTP sponsored doxylamine study at NCTR (Final Report 406 and 407).

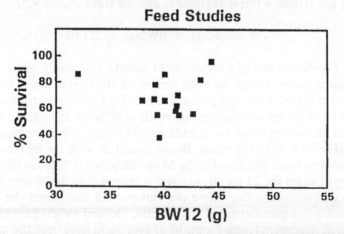

Figure 9.6.   BW12 and survival at 24 months. Body weight at 12 months on study and survival at 105 weeks (24 months) on study. Data obtained from the same experiments as in Figure 9.4.

evaluation of hormesis. This is especially true since the effect of calories on longevity is nonmonotonic. Too few calories result in death from malnutrition, while more than the amount required to survive appears to produce increased risk of death. The effect is so striking that toxicity testing using Sprague-Dawley rats is presently in a quandary because of the extremely poor survivals seen in long-term tests. In recent assays the males reach average body weights of 900–1000 grams,[14] with very poor survival at 24 months on test, a result which can invalidate the test. Because the underlying variable is nonmonotonic, and it significantly affects toxicity, hormesis has become a factor in almost every toxicity test.

Hormesis and Cancer

Although caloric restriction (CR) has an effect on longevity, even more significant are the effects on neoplasms. The inhibition of neoplasms by caloric restriction has been well documented in mice and rats since the original work of Tannenbaum in the 1940s (cited in References 4–6). Figure 9.7 shows some details of that effect in the F-344 rat. It can be seen that the effect of CR is to greatly delay the onset of pituitary tumors in males, with a

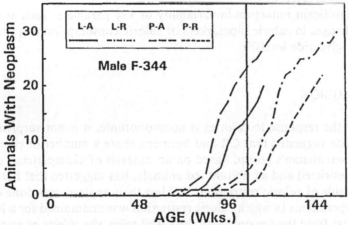

Figure 9.7. Age at death with leukemia and pituitary adenomas in F-344 male rats. Number of leukemia/pituitary adenomas and age of animals spontaneously dying. L-A is number of leukemias in ad libitum animals, L-R is leukemia in restricted animals, P-A is pituitary adenomas in ad libitum animals, P-R is pituitary adenomas in restricted animals. Vertical line is 111 weeks of age, approximately equivalent to the age of termination of an NTP bioassay. Note that pituitary adenomas are much more affected by caloric restriction than leukemias, so much so that leukemias are observed sooner in restricted animals than pituitary adenomas, the reverse of what occurs for ad libitum animals.

lesser effect on leukemias, resulting in a switch in the tumors which appear first in animals with dietary manipulation. A marker for an effect similar to that seen with caloric restriction is a differential effect on tumors, with endocrine tumors, such as pituitary tumors, being most inhibited.

Methods to Address the Question of Calories and Hormesis

Interpretation of the nonmonotonicity of dose-response of a toxicant in light of effects on changes related to caloric consumption is not a trivial task. Simply measuring consumption is complicated by the effects of wastage, variability, and activity. Activity changes have been seen with alteration of caloric consumption.[15] These changes are significant enough to result in alteration of blood corticosterone levels due to differential stimulation of restricted animals.[16] A measure, such as BW, can integrate some of the effects of activity, but becomes problematic when disease, which often results in either weight loss or weight gain, intervenes. A cyst can increase BW, but be unrelated to energy availability. Thorough analysis of weight changes, pathology, and food consumption are important to understanding caloric alteration as a mechanism for hormesis.

Characteristics which suggest a caloric mechanism include:

1. selective effects on endocrine organ tumors, a major target for caloric effects;
2. significant reduction in variability of key parameter such as BW;
3. changes in caloric biomarkers of calorie consumption (e.g., blood triglyceride levels[10]).

CONCLUSION

Since the response to calories is nonmonotonic, it is not surprising that Totter has suggested that CR and hormesis share a number of characteristics.[17] Boxenbaum's group, based on an analysis of Gompertzian parameters in restricted and ad libitum fed animals, has suggested that hormesis is not a result of caloric restriction,[18, 19] but their analysis was done on data from experiments in which caloric restriction was maintained for a lifetime. As is clear from the example with the LAF1 mice, the effects of an agent on growth, weight, etc., can be multiple, resulting in increases and decreases at different times. These changes will interact with the normal progression of the diseases, resulting in mortality in the individual strain, and the end result may not be simple. Lower consumption at one point in the lifespan may have different biological effects than at other points. Gompertzian parameter analysis does not readily accommodate factors such as these.

Mechanisms responsible for the actions of CR may be important in understanding basic mechanisms in hormesis. As the understanding of CR deepens, it may aid in deriving mechanisms for other nonmonotonic relationships that can be used to improve health and better the estimation of risk from exposure to toxic agents.

REFERENCES

1. Calabrese, E., M. McCarthy, and E. Kenyon. "The Occurrence of Chemically Induced Hormesis," *Health Physics*, 52:531 (1987).
2. Turturro, A., N.P. Singh, M.J.W. Chang, and R.W. Hart. "Effects of Transplacentally Administered Polyaromatic Hydrocarbons on the Genome of Developing and Adult Rats Measured by Sister Chromatid Exchanges," in *Polynuclear Aromatic Hydrocarbons: Physical and Biological Chemistry*, M. Cooke, A.J. Dennis, and G. Fischer, Eds., Battelle Press, Columbus, OH, 1982, p. 825.
3. Interagency Staff Group, "Chemical Carcinogens: A Review of the Science and Associated Principles," *Environ. Health Perspect.*, 67:201 (1986).
4. Allaben, W., M. Chou, R. Pegram, J. Leakey, R. Feuers, P. Duffy, A. Turturro, and R. Hart, "Modulation of Toxicity and Carcinogenesis by Caloric Restriction," *Korean J. Toxicol.*, 6:167 (1990).
5. Hart, R.W., J.E.A. Leakey, W.T. Allaben, M. Chou, P.H. Duffy, R.J. Feuers, and A. Turturro, "Role of Nutrition and Diet in Degenerative Processes," *Int. J. Toxicol. Environmental Hlth.*, 1:26 (1993).
6. Kritchevsky, D., and D. Klurfield. "Caloric Effects in Experimental Mammary Tumorigenesis, *Am. J. Clin. Nutr.*, 45:236 (1987).
7. Imai, K., S. Yoshimura, K. Hashimoto, and G. Boorman. "Effects of Dietary Restriction on Age-Associated Pathological Changes in Fischer 344 Rats," in *Biological Effects of Dietary Restriction*, L. Fishbein., Ed., Springer-Verlag, N.Y., 1991, p. 87.
8. Shimokawa, I., B.P. Yu, and E.J. Masoro. "Influence of Diet on Fatal Neoplastic Disease in Male Fisher 344 Rats," *Journal of Gerontology: Biological Sciences*, 46:B228 (1991).
9. Turturro A., and R. Hart. "Longevity-Assurance Mechanisms and Caloric Restriction," *Ann. New York Academy of Sciences*, 621:363 (1991).
10. Turturro A., and R. Hart. "Dietary Alteration in the Rate of Cancer and Aging," *Experimental Gerontology*, 27:583 (1992).
11. Sagan, L., "What is Hormesis?," *Health Physics*, 52:521 (1987).
12. National Toxicology Program, Carcinogenicity Studies of Methyl Bromide, NTP TR 385, 1992.
13. Congdon, C., "A Review of Low-Level Ionizing Radiation Studies in Mice and Guinea Pigs," *Health Physics*, 52:593 (1987).
14. Keenan, K., "The Effect of Diet and Caloric Optimization (Caloric Restriction) on Rat Survival and Carcinogenicity," *Proc. of the 43rd Meeting of the Amer. Coll. Vet. Path.*, 1992, p. 227.
15. Duffy, P., R. Feuers, J. Leakey, K. Nakamura, A. Turturro, and R. Hart. "Effect of Chronic Caloric Restriction on Physiological Variables That Modulate Energy Metabolism in the Male Fischer-344 Rat," *Mech. Age. Devel.*, 48:117 (1989).
16. Holson, R., P. Duffy, S. Ali, and F. Scalzo. "Aging, Dietary Restriction, and

Glucocorticoids: A Critical Review of the Glucocorticoid Hypothesis," in *Biological Effects of Dietary Restriction*, L. Fishbein, Ed., Springer-Verlag, N.Y., 1991, p. 123.

17. Totter, J., "Physiology of the Hormetic Effect," *Health Physics*, 52:549 (1987).

18. Neafsey, P., B. Boxenbaum, D. Ciraulo, and D. Fournier. "A Gompertz Age-Specific Mortality Rate Model of Aging, Hormesis and Toxicity: Fixed-Dose Studies," *Drug Metabol. Rev.*, 19:369 (1988).

19. Neafsey, P., B. Boxenbaum, D. Ciraulo, and D. Fournier. "A Gompertz Age-Specific Mortality Rate Model of Aging: Modification by Dietary Restriction in Rats," *Drug Metabol. Rev.*, 21:351 (1989).

PART III

Biological Effects of Low Level Exposure to Radiation

CHAPTER 10

What Can Be Learned from Epidemiologic Studies of Persons Exposed to Low Doses of Radiation?

Ethel S. Gilbert, Pacific Northwest Laboratory, Richland, Washington

INTRODUCTION

The main objective of radiation risk assessment is to determine the risk of various adverse health effects associated with exposure to low doses and low dose rates. Extrapolation of risks from studies of persons exposed at high doses (generally exceeding 1 Sv*) and dose rates has been the primary approach used to achieve this objective. The study of Japanese atomic bomb survivors in Hiroshima and Nagasaki has played an especially important role in risk assessment efforts.

A direct assessment of the dose-response function based on studies of persons exposed at low doses and dose rates is obviously desirable. Several studies of workers involved in the production of both defense materials and nuclear power have been conducted in the United States, Great Britain, and Canada, and further studies of nuclear power workers in several countries are being planned. In addition, the Japanese atomic bomb survivor study includes thousands of survivors exposed at low doses.

Table 10.1 shows the dose distributions both for the Japanese atomic bomb survivors and for nuclear workers at the Hanford site, a population that is typical of many of the worker studies. Unlike the worker study, the atomic bomb survivor study includes many persons exposed to doses exceeding 1 Sv, and these large doses tend to dominate dose-response analyses. For Hanford workers, the doses shown in Table 10.1 are cumulative doses (added up over all years of employment); annual doses rarely exceed

---

*Exposure in this paper is expressed in sievert (Sv) or millisieverts (mSv). A sievert is numerically equivalent to the absorbed dose in gray (Gy) multiplied by quality factors expressing the biologic effectiveness of the radiation type. Sievert and gray are the current internationally accepted units of measurement of dose equivalent and dose, respectively. Because some readers may be more familiar with the older units rem and rad, it is noted that 10 mSv = 1 rem and 10 mGy = 1 rad.

Table 10.1. Number of Subjects (Proportion) by Total Radiation Dose for Japanese
Atomic Bomb Survivors and for Hanford Workers Employed at Least Six
Months and Monitored for External Radiation

Dose Category (Gy[a])	Japanese Atomic Bomb Survivors[b]		Hanford Workers[c]	
<0.01	34,272	(0.451)	21,530	(0.660)
0.01–	19,192	(0.253)	7,396	(0.227)
0.05–	4,129	(0.054)	1,559	(0.048)
0.10–	5,172	(0.068)	1,058	(0.032)
0.20–	6,558	(0.086)	955	(0.029)
0.50–	3,616	(0.048)	145	(0.004)
1.00–	1,946	(0.026)	0	
2.00–	637	(0.008)	0	
3.00–	211	(0.003)	0	
4.00+	258	(0.003)	0	
Total	75,991	(1.000)	32,643	(1.000)

[a]For Hanford workers, doses are in Sv, but because most exposure was to photons, the dose in Gy and dose equivalent in Sv was similar for most workers.
[b]From Table AII, Shimizu et al.[26]
[c]For Hanford workers, the dose is cumulative dose from 1944 through 1985.

50 mSv. In comparing these dose distributions, it should be kept in mind that dose estimates for workers, which have been obtained from personal dosimeters worn by the workers, are probably more precise than those of atomic bomb survivors.

For risk assessment purposes, interest usually centers on doses less than 0.1 Sv. For assessing risks of environmental exposures (for example, world-wide exposures resulting from the Chernobyl accident), doses that are orders of magnitude lower than 0.1 Sv are of interest.

This chapter focuses on the potential of both current and future nuclear workers studies for investigating the dose-response function at low doses, and also discusses analyses making use of the low dose portion of the atomic bomb survivor data. Difficulties in using these data are the statistical imprecision of estimated dose-response parameters, and potential bias resulting from confounding factors and from uncertainties in dose estimates.

LOW DOSE ASSESSMENT BASED ON THE JAPANESE ATOMIC BOMB SURVIVOR DATA

Risk estimates provided by the National Research Council's Report on the Health Effects of Exposure to Low Levels of Ionizing Radiation (BEIR V),[1] and by the 1990 Recommendations of the International Commission on Radiological Protection (ICRP),[2] were based on Japanese atomic bomb survivor data using doses ranging from 0 to 4 Sv. The observed effects in the high dose portion of these data (greater than 1 Sv) dominated results of these analyses. However, the majority of atomic bomb survivors had relatively low doses (Table 10.1), and various efforts have been made to use the low dose portion of the data to provide a direct assessment of low dose risks. The most recent of these efforts by Shimizu et al.[3] and by Vaeth et al.[4] are briefly reviewed below.

Shimizu et al. conducted analyses of mortality data (and incidence data for breast and thyroid cancer) to investigate the possibility of a protective effect of radiation in the low dose range. For leukemia, the relative risks for the dose categories 0.01 to 0.019, 0.02 to 0.049, 0.05 to 0.099 (relative to a baseline category of 0.00 to 0.01 Sv) were all less than one, but confidence limits included one. Fitting linear, linear-quadratic, and quadratic models to the < 0.05 Sv range indicated that both positive and negative slopes in the low dose-range (< 0.01 Sv) were consistent with the data.

For cancers other than leukemia, a linear model fit the data well, and there was no evidence that the slope in the lower dose portion of the curve differed from that based on the entire dose range. Specific analyses of stomach cancer, lung cancer, breast cancer, and thyroid cancer also failed to yield evidence of departure from linearity. Incidence data were used for analyses of breast and thyroid cancer.

Vaeth et al.[4] conducted analyses of incidence data on leukemia and all cancers except leukemia. Within the context of the linear-quadratic model, they assessed the maximum curvature consistent with the data by estimating the low-dose extrapolation factor (LDEF), where the LDEF is defined as the value that the linear risk estimate based on high-dose data would need to be divided by to obtain risks at low doses. For cancers other than leukemia, the best estimate of the LDEF was one, with 90% confidence limits (0.8, 1.5). For leukemia, however, the estimate was 2.5 with 90% confidence limits (1.4, 9), providing evidence of departure from linearity over the entire dose range. Furthermore, results for leukemia were not consistent with those for cancers other than leukemia, indicating that the shape of the dose-response at low doses may vary by cancer type.

Limitations in dosimetry for distinguishing among doses at very low levels may have limited the potential of the analyses described above. Vaeth et al.[4] evaluated the possible impact of random dosimetry errors on their results, and found that this did not greatly modify conclusions. However, neither the analyses by Shimizu et al.[3] nor by Vaeth et al. accounted for the possibility of special uncertainties in low estimated doses. Many of the survivors with low doses were located far from the hypocenters, and thus did not have detailed shielding histories. Dose estimates for these survivors may thus have been subject to larger relative errors than dose estimates for survivors for whom shielding information was available.

THE HANFORD WORKER STUDY

The most promising low-dose epidemiologic studies for direct evaluation of health effects resulting from low level exposures are those of workers who have been exposed occupationally to low levels of radiation. In this section, data from a study of workers at the Hanford site are used to illustrate problems and limitations in interpreting data from occupationally exposed populations. The Hanford site, which is located in southeastern Washington state, was established in the 1940s as an installation for plutonium production, although since that time, efforts have been expanded to include power production and a variety of research activities. The study population includes more than 44,000 employees of U.S. Department of Energy contractors, making it one of the largest worker populations under study. Recent dose-response analyses have been limited to 32,643 workers who were employed at the Hanford site for at least six months and who were monitored for external radiation.

From Table 10.1, it can be seen that for the majority of workers, the total occupational dose was small, less than 10 mSv. Most of these workers did not perform radiation work but were nevertheless monitored for external radiation exposure. Only about 10% of Hanford workers accumulated doses of 50 mSv or more, and most of the exposure was received by male workers.

A description of the statistical methods used to analyze the Hanford data, and detailed results of analyses are presented by Gilbert et al.[5] Here, only a few example results are presented.

Figure 10.1 shows results of dose-response analyses of mortality from all cancers except leukemia. Estimates of the relative risks (with 90% confidence limits) are shown for five categories defined by cumulative dose. The category < 10 mSv serves as the baseline, and a minimal latency period of 10 years was allowed for by lagging doses by 10 years. In addition, the best fitting linear function is shown, based on the assumption that the relative risk was of the form $1 + \beta z$, where z is cumulative dose, and β is the excess relative risk. The estimated excess relative risk was based on ungrouped doses, not the five categories for which relative risks are presented.

Figure 10.1 also shows the linear excess relative risk estimate for Japanese atomic bomb survivors exposed in adulthood, obtained from analyses presented in a report of the United National Scientific Committee on the Effects of Atomic Radiation (UNSCEAR[6]). The UNSCEAR estimates served as the basis for lifetime risk estimates presented by the ICRP,[2] although the ICRP recommended reducing these estimates by a factor of two for exposures received at low doses and dose rates.

It is clear from Figure 10.1 that the Hanford data are consistent with no effect, with protective effects, and with effects that are several times larger than linear estimates obtained from high-dose data. The estimated excess relative risk for Hanford was -0.15% per 10 mSv with 90% confidence

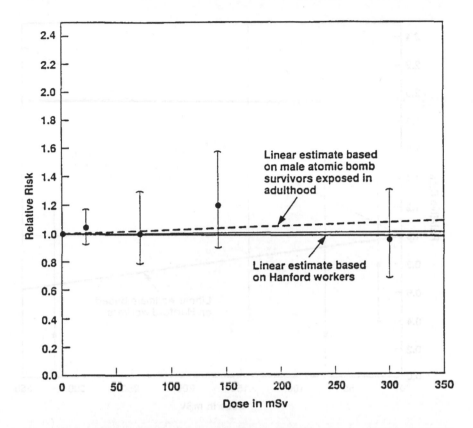

Figure 10.1. Relative risk estimates with 90% confidence limits for all cancer except leukemia, based on Hanford workers employed at least six months and monitored for external radiation.

limits ($<$ 0, 1.0% per 10 mSv) compared with the UNSCEAR linear estimate of 0.24% for males exposed in adulthood.

With so little evidence of *any* type of dose-response, investigating the shape of the dose-response curve is not likely to be fruitful. Nevertheless, attempts were made to fit a model in which the relative risk was of the form $1 + \beta z^{\gamma}$. For γ ranging from 0.01 to 20, the likelihood ratio chi-square ranged only from 0.0 to about 0.5, indicating that these data have almost no information on the shape of the dose-response. The excess relative risk was also estimated based on only doses less than K mSv with K taken to be 200, 100, 50, and 10. In all cases, 90% confidence intervals included zero, and upper confidence limits increased as K decreased.

Figure 10.2 shows analyses of leukemia mortality (excluding chronic lymphatic leukemia); in this case, doses were lagged for two rather than 10 years. The excess relative risk for Hanford was –1.1% per 10 mSv with 90% confidence limits ($<$ 0 1.9% per 10 mSv) compared with the UNSCEAR

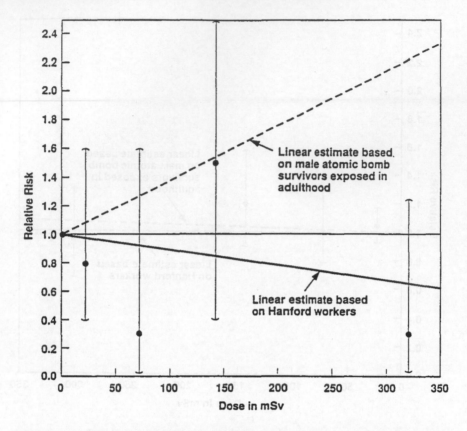

Figure 10.2. Relative risk estimates with 90% confidence limits for leukemia (excluding chronic lymphatic leukemia), based on Hanford workers employed at least six months and monitored for external radiation.

linear estimate of 3.7% for males exposed in adulthood. Thus, Hanford leukemia results are consistent with no risk, with protective effects, and with positive effects. The upper limit of 1.9% can be roughly interpreted as a value such that larger risk estimates can be rejected at the 0.05 level, indicating that the ICRP recommendations probably have not seriously underestimated risks. However, obviously it must be kept in mind that confidence limits reflect only uncertainty resulting from random variation, and do not reflect uncertainties from other biases to which worker-based risk estimates are subject.

OTHER STUDIES OF WORKERS EXPOSED TO LOW DOSES OF EXTERNAL RADIATION

Hanford is one of several worker studies that have reported results.[7] In the United States, results of dose-response analyses on workers exposed to external radiation have also been reported for workers at Oak Ridge National Laboratory (ORNL)[8] and Rocky Flats Weapons Plant.[9] In the United Kingdom, results have been reported for workers at the Atomic Energy Authority,[10] the Sellafield plant of British Nuclear Fuels,[11] and the Atomic Weapons Establishment.[12] Results of a study of the United Kingdom National Registry for Radiation Workers (NRRW)[13] have also been reported, and includes virtually all of the workers in the three individual studies. In Canada, a study of workers at Atomic Energy of Canada Limited (AECL)[14] has reported results.

Table 10.2 shows risk estimates for Hanford, Oak Ridge National Laboratory (ORNL), Rocky Flats, and the National Registry Radiation Worker study in the United Kingdom. As noted above, the latter study includes most workers in other United Kingdom studies, and at the current time represents the best available summary of these data.

For all cancer, risk estimates were positive in all studies but Hanford. However, the NRRW confidence intervals indicate consistency with no effect and with estimates that have been recommended by the ICRP. The ORNL estimate was several times larger than estimates based on atomic bomb survivors. Wing et al.[8] did not present confidence limits for this estimate, but Gilbert[15] provided an alternative analysis which yielded a risk estimate of 2.9% per 10 mSv with a 90% confidence interval of (0.4%, 6.5%). The lower limit was still larger than estimates based on atomic bomb survivors, but bias may have resulted because of differential smoking patterns, and Gilbert did not have data available to adjust for differential socioeconomic status. For leukemia, NRRW data indicate that the possibility of no risk or protective effects are unlikely. Although estimates were larger than estimates that form the basis of ICRP recommendations, confidence limits indicate consistency with such estimates.

In addition to the individual studies, there are national and international efforts to combine data from these studies. Results of combined analyses of the data from the three U.S. studies have been reported[16] (and are currently being updated), and results of combined analyses of data from the three U.K. studies[17] are expected to be reported soon. A collaborative effort to

Table 10.2. Excess Relative Risk Estimates with 90% Confidence Intervals for All Cancers and for Leukemia (Excluding Chronic Lymphatic Leukemia). Based on a 10-Year Lag for All Cancers and on a 2-Year Lag for Leukemia

| Population | Excess Relative Risk Estimates (% per 10 mSv) | | Total Person–Sv |
	All Cancer	Leukemia	
Hanford Site[5]	−0.1% (<0%, 0.0%)	−1.1% (<0%, 1.9%)	854
Oak Ridge National Laboratory[8]	3.3%		144
Rocky Flats Weapons Plant[16]	<0% (<0%, 2.8%)	4.3% (<0%, 52%)	241
National Registry for Radiation Workers[13]	0.5% (<0%, 1.2%)	4.3% (0.4%, 14%)	3200

conduct combined analyses of the data from all three countries is being coordinated at the International Agency for Research on Cancer (IARC),[18] and a report describing these results is in preparation. In addition, a collaborative study of additional nuclear workers in several countries is being planned, with IARC again serving as the coordinating agency.[19]

An assessment of the potential of future studies is of interest, but this assessment requires detailed data on the magnitude of the doses and the time and age that these doses were received; data of this type are not currently available. However, by using data that *were* available and making assumptions where data were not available, Cardis and Esteve[19] attempted to assess this potential. They estimated that if linear estimates recommended by the BEIR V Committee[1] were correct, then 95% confidence limits based on the combined worker data would be about ±70% for all cancer except leukemia, and about ±60% for leukemia. They also estimated that if effects were reduced by a factor of two over those provided by BEIR V, then the upper limit based on the combined data would be about 20% higher than the BEIR V linear estimate; while if effects were reduced by a factor of five, the upper limit would be about 90% of the BEIR V estimate. At the current time, most of the information comes from countries with older nuclear power industries, particularly France, Japan, and the United Kingdom.

The assessment above did not consider all sources of sampling variability and, as noted above, was necessarily based on many arbitrary assumptions about dose and age distributions. On the other hand, the estimates were based on mortality data through 1991; if the studies were extended beyond this period, additional information would result.

It is also important to note that in many risk assessment situations, interest centers on extremely low doses, possibly at the 1 mSv or lower level. It must be recognized that worker-based estimates are driven by much larger doses, generally greater than 100 mSv (0.1 Sv), and extrapolation would still be required to obtain estimates at extremely low doses. In addition, the confidence limits presented in Table 10.2, and the projections discussed above did not include uncertainty that may have resulted from confounding that has not been accounted for in the analysis, or from biases and uncertainties in the dose estimates that have been used. These problems are discussed below.

BIAS RESULTING FROM CONFOUNDING

Epidemiologic studies are not randomized experiments, and cancer risks may differ by level of exposure for reasons other than the exposure itself. This difficulty is particularly troublesome in attempting to assess very small changes in risk such as those likely to be encountered in worker studies. Note, for example, that based on linear extrapolation from high dose data, the relative risk for a male worker exposed to 0.2 Sv is estimated to be about 1.4 for leukemia and about 1.025 for cancers other than leukemia. Even if sufficient data could be accumulated to obtain estimates that were statistically very precise, it is almost never possible to rule out the possibility that unidentified confounders might have biased results by 10% to 20%, and this could limit the value of analyses based on such data.

In worker studies, those with higher doses tend to have been employed longer in the industry, and may also differ in their socioeconomic characteristics from workers with lower doses. In Hanford workers, both factors have been shown to be linked to cancer mortality. For example, in analyses presented in Gilbert,[20] cancer death rates were shown to decrease with increasing length of employment.

Recently, job category data for Hanford workers were used to define a socioeconomic variable that is similar to the social class variable used in some worker studies in the United Kingdom. Using the professional and technical category as a baseline, relative risks for cancer mortality (with 90% confidence limits) were 1.1 (0.9, 1.3) for clerical workers, 1.3 (1.1, 1.4) for skilled and semiskilled workers, and 1.5 (1.2, 1.8) for unskilled workers. Average doses for the four categories were 17.7, 3.6, 53.0, and 5.2 mSv, respectively.

Results of dose-response analyses shown in Figures 10.1 and 10.2 and discussed above were adjusted for both length of employment (using two strata < 5 years and 5+ years), and socioeconomic status (using four strata). It is important to note that all strata defined by length of employment and socioeconomic status included substantial numbers of workers with little or no exposure, so that such stratification did not seriously reduce the statistical information available for dose-response analyses.

Table 10.3 shows risk estimates based on alternative choices of controlling factors. For cancers other than leukemia, not adjusting for socioeconomic status increased risk estimates and upper confidence limits, while not adjusting for length of employment had the opposite effect. For leukemia, the alternative choices all increased the upper confidence limits. There was

Table 10.3. **Excess Relative Risk Estimates (with 90% Confidence Limits) for Leukemia and All Cancer Except Leukemia Based on Alternative Choices of Controlling Factors. Expressed as Percent Increase per 10 mSv.**

Additional Controlling Factors[a]	All Cancer Except Leukemia	Leukemia
Number of years monitored and socioeconomic status	−0.04% (< 0, 1.0%)	−1.1% (<0, 1.9%)
Number of years monitored	0.50% (< 0, 1.6%)	−0.89% (<0, 3.1%)
Socioeconomic status	−0.28% (< 0, 0.6%)	−0.80% (<0, 3.4%)
None	0.15% (< 0, 1.1%)	−0.71% (<0, 3.6%)

[a]All analyses were controlled for age, calendar year, and sex.

little evidence of an association of leukemia mortality with either length of employment or socioeconomic status, but this may be because of limited sample size.

Although the alternative results presented in Table 10.3 do not alter the general conclusion that risks are consistent with estimates obtained from high dose data, they *do* impact more precise conclusions about the exact magnitude of departure from current estimates that would be consistent with these data. The decision to stratify on both length of employment and socioeconomic status is necessarily somewhat arbitrary. Clearly, these variables are not causal factors in themselves, but rather are surrogates for other factors such as smoking and diet, for which we do not have data. The variation in results seen in Table 10.3 and the absence of adequate data on all factors that might affect cancer risks indicates additional uncertainty in worker-based risk estimates that is not included in usual statistical confidence limits.

BIAS RESULTING FROM DOSIMETRY UNCERTAINTIES

The availability of objective quantitative measurements of exposure to external radiation is a major strength of both the atomic bomb survivor study and the nuclear worker studies. However, the dose estimates used in these studies are subject to several sources of bias and uncertainty, which have not usually been taken into account in dose-response analyses. For the atomic bomb survivors, sources of uncertainty include uncertainty in the yield of the bombs (especially the Hiroshima bomb), uncertainty in the air dose curves (the manner that radiation was transported), uncertainty in shielding factors for those survivors who were in buildings or otherwise shielded, and uncertainty concerning the information provided by survivors on location and shielding at the time of the bombs. For nuclear workers, dose estimates are obtained from personal dosimeters worn by the workers. Sources of uncertainty in these estimates include random variation in laboratory measurements, failure of the dosimeter to respond accurately to all radiation energies to which workers are exposed, and failure of the dosimeter to respond accurately to radiation coming from all directions.

Bias in dose estimates obviously can bias risk estimates expressed per unit of dose. In addition, strictly random error in dose estimates is known to bias estimated regression coefficients toward the null, and may also result in the underestimation of uncertainty and in distortion of the shape of the dose-response curve.[21-23] In general, the amount of distortion resulting from dose measurement errors depends on the magnitude of the error relative to the overall variability in the dose distribution. This relative error is likely to be especially large for analyses restricted to low doses, and may thus greatly limit the amount that can be learned from low-dose data.

An objective of analyses of worker data is to compare risk estimates with those that have been obtained from high-dose data, and thus to evaluate the extent to which risk estimates might be reduced (or increased) over those obtained through extrapolation. Risk estimates based on high-dose data such as the atomic bomb survivors have usually been based on absorbed dose to various organs of the body. By contrast, the dose estimates used in nuclear worker studies were those obtained for radiation protection purposes, and probably overestimate organ doses in many cases. The extent of this overestimation depends on the energy and type of radiation to which workers were exposed, and also on the direction from which the radiation was received. In addition, dosimetry practices at most facilities changed over time as technology improved so that the extent of the bias may also depend on the time the dose was received. Because of uncertainty in the exposure situations of workers, and because of uncertainty in dosimetry practices during early years of operation, factors for converting recorded doses to doses to various organs of the body cannot be estimated precisely.

The relative uncertainty in the measurement of doses near background levels is particularly large. For example, in an investigation of film dosimetry in atmospheric nuclear tests,[24] the uncertainties introduced in laboratory processing and calibration of film dosimeters for many of the test series evaluated were assessed by assigning a standard error that was 10% of the estimated exposure for exposures above 2 mSv. However, this standard error was judged to be about 20%, 50%, and 100% for readings of 1, 0.4, and 0.2 mSv, respectively. This means that exposures below about 0.4 mSv could not be statistically distinguished from zero.

In some facilities, measured doses that were below some "threshold" level were recorded as zero. Bias related to this practice was most likely to have been a problem in the 1940s and 1950s when dosimeters in most facilities were exchanged weekly or biweekly, and in which the "threshold" may have been set at a fairly high level. For example, prior to 1961, dosimeters for United Kingdom Atomic Energy Authority workers were exchanged weekly, and readings below 0.5 mSv were recorded as zeros.[25] Such a procedure does not appear to have been applied at Hanford.

An additional problem in estimating very low doses occurs because of the need for adjustment for background exposure, which is accomplished by subtracting readings from control dosimeters. For personnel with very little or no occupational dose, this subtraction can yield negative estimates and these estimates are recorded as zeros. Because positive results for unexposed workers are recorded and are not compensated by the subtraction of negative results, cumulative doses for workers with little or no occupational dose tend to be overestimated. It is noted that at most U.S. facilities, including Hanford, even personnel in facilities where no radiation work is performed are monitored for external radiation exposure.

In an effort to gain information on Hanford practices for handling very low doses, we examined the distributions of recorded doses for each calendar year for workers whose job titles indicated very little likelihood of true occupational exposure. Over the period 1944 through 1978, the percentage of these workers with annual doses recorded as zero ranged from 5% to 89%, the median annual recorded doses ranged from 0 to 2.9 mSv, while the 90th percentiles ranged from 0.2 mSv to 5.6 mSv. These ranges indicate that procedures for handling very low measurements varied considerably from year to year. Because these procedures are not always well documented, especially for earlier years, it is difficult to know what sort of correction (if any) should be made.

Sensitivity analyses based on the Hanford data have shown that various treatments of very low doses did not have much effect on overall dose-response analyses. However, analyses by Inskip et al.[25] of the U.K. Atomic Energy Authority worker data indicated that the dose-response was modified by adjustment for the practice of recording doses below a threshold value as zero. It is clear that dosimetry problems in measuring very low doses make it nearly impossible to provide a meaningful investigation of dose-response in the less than 10 mSv range.

Other problems in measuring doses received by nuclear workers include underestimation of dose from neutrons prior to about 1970, and the inability to accurately measure doses from internal depositions of radionuclides. Because of these problems, the protocol for the international combined analyses[18] requires identifying workers with potential for doses from these sources, and some analyses will be conducted that exclude these workers. Additional dosimetry problems are incomplete information on occupational exposure received at facilities other than the one under study, and lack of data on medical and background exposures.

Procedures for accounting for biases and uncertainties in dose estimates are available. However, these procedures are often very difficult to apply, and require knowledge of the nature and magnitude of dose measurement errors, including knowledge of the extent to which errors are correlated between subjects. It is important to attempt to apply such methods, but it will probably be necessary to base these applications on overly simple assumptions, and to consider such analyses as sensitivity analyses, conducting analyses based on several alternative assumptions. The end result of such analyses will *not* be to eliminate bias and uncertainty resulting from imperfect dosimetry, but rather to provide a better assessment of the additional uncertainty in the estimated dose-response resulting from this problem.

CONCLUSIONS

The low-dose portion of the Japanese atomic bomb survivor data provides some direct information on effects at low doses. For leukemia, these data indicate that linear extrapolation is likely to overestimate risks, and that the dose and dose rate effectiveness factor (DDREF)* of 2 recommended by the ICRP is a reasonable choice, although the data were also consistent with a range of values from 1.4 to 8 (90% confidence limits). For cancers other than leukemia, a linear function provided a good fit over both the high and low dose portions of the data. All exposure in this population was received at high dose rates, and thus these data cannot be used to assess the modifying effects of dose rate.

Analyses of the low-dose portion of the atomic bomb survivor data indicated that the potential of these data for learning about the specific dose response in the less than 0.1 Sv range was extremely limited. Just considering sampling variation, the possibility of no effect or of protective effects at doses at these levels could not be ruled out. Consideration of dosimetry uncertainties and confounding provide additional limitations.

Data from several studies of nuclear workers who have been exposed occupationally to radiation provide the most promising direct evaluation of health effects resulting from exposure at low doses and dose rates. Based on results of these studies that have been reported thus far, risk estimates from most studies are consistent with risk estimates obtained through extrapolation, and risks as high as an order of magnitude larger appear to be extremely unlikely. However, confidence limits also indicate consistency with estimates that are a few times larger than those obtained through extrapolation and with the possibility of no effect or of a protective effect.

Confidence limits that have been reported did not include uncertainty resulting from possible confounding or from bias and uncertainties in the dose estimates. To provide a more rigorous comparison of worker-based estimates with those from other sources, it is important to conduct analyses that attempt to incorporate dosimetry uncertainties, although it is recognized that this cannot be accomplished perfectly and will necessarily involve subjective judgments. A subjective assessment of uncertainty from potential confounding could also be made and incorporated into overall "credibility" intervals.

Combined international analyses and additional data from future nuclear worker studies will reduce the statistical uncertainty in risk estimates. Combining data from several countries and facilities may also result in some cancelling-out of bias related to confounding, but this cannot be guaranteed, and to the extent that bias is similar from study to study, potential confounding will remain a problem. More recent dosimetry and better documentation can be expected to reduce, but not eliminate, uncertainties related to dosimetry compared to studies that have already reported results. In spite of these limitations, worker studies provide an extremely useful check on the validity of estimates obtained through extrapolation from

*The dose and dose rate effectiveness factor (DDREF) is the factor needed to reduce effects obtained through linear extrapolation from high dose/high dose rate data to obtain estimates of risks at low doses and dose rates. The value of the DDREF currently recommended by the ICRP is two, a value that was obtained primarily through consideration of experimental data.

studies of persons exposed at high doses and dose rates. If risks have been seriously underestimated by the extrapolation process, these studies have adequate power to detect this, while if risks have not been seriously underestimated, then these studies can provide useful upper limits on risks based on a direct assessment at low doses and dose rates.

However, epidemiologic data, whether from low-dose atomic bomb survivors or from workers, have not been informative regarding the possibility of protective effects or thresholds at very low doses (less than 10 mSv), and have failed to yield precise estimates of the DDREF. Sampling variation alone severely limits low-dose studies, but even if sample size could be greatly increased, uncertainty resulting from potential confounding and from dosimetry limit the information that epidemiologic studies can provide on these issues.

ACKNOWLEDGMENT

This work was supported by the U.S. Department of Energy under Contract DE-AC06–76RLO-1830.

REFERENCES

1. National Academy of Sciences, *Health Effects of Exposure to Low Levels of Ionizing Radiation, BEIR V. Report of the Committee on the Biological Effects of Ionizing Radiation*, National Research Council. National Academy of Sciences, Washington DC, 1990.
2. International Commission of Radiological Protection (ICRP). *1990 Recommendations of the International Commission on Radiological Protection*, Publication 60. *Ann. ICRP*. 21:1 (1991).
3. Shimizu. Y., H. Kato, W.J. Schull, and K. Mabuchi, "Dose-Response Analysis of Atomic Bomb Survivors Exposed to Low-Level Radiation," *RERF Update*. 4(3):3 (1992).
4. Vaeth, M., D.L. Preston, and K. Mabuchi, "Extrapolating Life Span Study Cancer Risk Estimates to Low-Dose Radiation Exposures," *RERF Update*, 4(3):5 (1992).
5. Gilbert, E.S., E. Omohundro, J.A. Buchanan, and N.A. Holter. "Mortality of Workers at the Hanford Site: 1945–1986," *Health Phys.* 64:577–590 (1993).
6. *Sources, Effects, and Risks of Ionizing Radiation*, United Nations Scientific Committee on the Effects of Atomic Radiation, 1988 Report to the General Assembly, with Annexes, United Nations, New York, 1988.
7. Gilbert, E.S., "Studies of Workers Exposed to Low Doses of External Radiation." in *Occupational Medicine: State of the Art Reviews*. Vol. 6, No. 4, October-December 1991. Hanley & Belfus, Inc., Philadelphia, PA.
8. Wing, S., C.M. Shy, J.L. Wood, S. Wolf, D.L. Cragle, and E.L. Frome. "Mortality Among Workers at Oak Ridge National Laboratory: Evidence of Radiation Effects in Follow-Up Through 1984," *JAMA*, 265:1397 (1991).

9. Wilkinson, G.S., G.L. Tietjen, L.D. Wiggs, W.A. Galke, J.F. Acquavella, M. Reyes, G.L. Voelz, and R.J. Waxweiler. "Mortality Among Plutonium and Other Radiation Workers at a Plutonium Weapons Factory," *Am. J. Epidemiol.* 125:231 (1987).

10. Beral, V., H. Inskip, P. Fraser, M. Booth, D. Coleman, and G. Rose. "Mortality of Employees of the United Kingdom Atomic Energy Authority, 1946-1979." *Br. Med. J.* 291:440 (1985).

11. Smith, P.G. and A.J. Douglas. "Mortality of Workers at the Sellafield Plant of British Nuclear Fuels." *Br. Med. J.* 293:845 (1986).

12. Beral, V., P. Fraser, L. Carpenter, M. Booth, A. Brown, and G. Rose, "Mortality of Employees of the Atomic Weapons Establishment, 1951-1982," *Br. Med. J.* 297:757 (1988).

13. Kendall, G.M., C.R. Muirhead, B.H. MacGibbon, J.A. O'Hagan, A.J. Conquest, A.A. Goodill, B.K. Butland, T.P. Fell, D.A. Jackson, M.A. Webb, R.G.E. Haylock, J.M. Thomas, and T.J. Silk, "Mortality and Occupational Exposure to Radiation: First Analysis of the National Registry for Radiation Workers," *Br. Med. J.* 304:220 (1992).

14. Howe, G.R., J.L. Weeks, A.B. Miller, A.M. Chiarelli and J. Etezadi-Amoli, "A Study of the Health of the Employees of Atomic Energy of Canada Limited. IV. Analysis of Mortality During the Period 1950-1981," AECL-9442. Pinawa, Manitoba, Atomic Energy of Canada Limited, 1987.

15. Gilbert, E.S., "Mortality of Workers at the Oak Ridge National Laboratory," (letter). *Health Phys.* 62:260 (1991).

16. Gilbert, E.S., S.A. Fry, L.D. Wiggs, G.L. Voelz, D.L. Cragle, and G.R. Petersen. "Analyses of Combined Mortality Data on Workers at the Hanford Site. Oak Ridge National Laboratory, and Rocky Flats Nuclear Weapons Plant," *Radiat. Res.* 120:19 (1989).

17. Carpenter, L., A. Douglas, C. Higgins, P. Fraser, V. Beral, and P. Smith. "Preliminary Results from the UK Nuclear Industry Combined Epidemiological Analysis of Mortality, 1946-88," in *International Conference on Effects of Low Dose Ionising Radiation: Implications for Human Health*, British Nuclear Energy Society, London, 1992.

18. Cardis, E., and J. Kaldor. *Protocol of Combined Analyses of Cancer Mortality Among Nuclear Industry Workers*, Internal Report No 89/005, International Agency for Research on Cancer (IARC), Lyon, France, 1989.

19. Cardis, E., and J. Esteve. *International Collaborative Study of Cancer Risk Among Nuclear Industry Workers. I-Report of the Feasibility Study*, Internal Report No. 92/001, International Agency for Research on Cancer (IARC), Lyon, France, 1992.

20. Gilbert, E.S., "Issues in Analyzing the Effects of Occupational Exposure to Low Levels of Radiation," *Statist. in Med.* 8:173 (1989).

21. Armstrong, B., "The Effects of Measurement Errors on Relative Risk Regressions." *Am. J. Epidemiol.* 132:1176 (1990).

22. Cochran, W.G., "Errors of Measurement in Statistics." *Technometrics.* 10:637 (1965).

23. Gilbert, E.S., "Accounting for Bias and Uncertainty Resulting from Dose Measurement Errors and Other Factors," in *The Future of Radiation Research*, Schloss Elmau, Federal Republic of Germany, March 4-8, 1991.

24. National Research Council, Committee on Film Badge Dosimetry in Atmo-

spheric Nuclear Tests. *Film Badge Dosimetry in Atmospheric Nuclear Tests*, National Academy Press, Washington DC, 1989.

25. Inskip, H., V. Beral, P. Fraser, M. Booth, D. Coleman, and A. Brown. "Further Assessment of the Effects of Occupational Radiation Exposure in the United Kingdom Atomic Energy Authority Mortality Study," *Br. J. Ind. Med.* 44:149 (1987).

26. Shimizu, Y., H. Kato, and W.J. Schull, "Studies of Mortality of A-Bomb Survivors. 9. Mortality, 1950-1985: Part 2. Cancer Mortality Based on the Recently Revised Doses (DS68)," *Radiat. Res.* 121:120 (1990).

CHAPTER 11

Positive Health Effects of Low Level Radiation in Human Populations

Myron Pollycove, United States Nuclear Regulatory Commission, Washington, DC

Increased longevity and decreased cancer death rates have been observed in populations exposed to high natural background radiation and reported for several decades.[1-7] These observations contradict the radiation paradigm that all radiation, including that of natural background, is harmful in linear proportion to low level dose. Such observations have been considered by recognized radiation scientists to be spurious or inconclusive because of unreliable public health data or undetermined confounding factors such as pollution of air, water and food, smoking, income, education, medical care, population density, and other socioeconomic variables. Attempts to establish a threshold level below which radiation is not harmful have been negated by the great difficulty of obtaining accurate data on large human populations required to demonstrate the absence of very low risks of low-level radiation predicted by linear extrapolation of high-level radiation health effects. During the past four years, however, several epidemiologic studies have demonstrated that exposure to low or intermediate levels of radiation have apparently resulted in positive health effects.

Low Level Radiation of Nuclear Shipyard Workers

A ten-year study by the Johns Hopkins Department of Epidemiology, School of Public Health and Hygiene, of nuclear shipyard workers was concluded recently.[8] The Technical Advisory Panel (TAP), chaired by Arthur C. Upton, advised on the research and reviewed results. John Cameron, a member of the TAP, summarized the study and stated, "This study is probably the best evidence that low levels of ionizing radiation are without health hazard."

The results contradict the conclusions of the BEIR V report that small amounts of radiation have risk—the linear risk hypothesis.[9] The database of almost 700,000 shipyard workers included almost 108,000 nuclear workers with exposures beginning in the 1960s until the end of 1981. Three study groups were selected: 33,352 non-nuclear workers (NNW), 10,462 nuclear workers with a working lifetime dose equivalent (DE) of less than 5mSv (NW$_{<5}$), and 28,542 nuclear workers with a DE greater than or equal to 5mSv (NW$_{>5}$). Five mSv (0.5rem) is approximately equal to the sea-level background radiation (340 mr/yr) one would receive in 1½ years. Deaths in each group were classified as due to: all causes, leukemia, lymphatic and hematopoietic cancers (LHC), mesothelioma, and lung cancer. The only cancer that showed a significantly increased incidence in the exposed groups as well as the NNW was the rare malignancy mesothelioma (36 deaths), a marker for asbestos exposure that is also associated with lung cancer. The data are summarized in Table 11.1.

The nuclear worker groups had a lower death rate from all causes, leukemia, and LHC than the non-nuclear workers. These apparently beneficial effects of low dose irradiation are consistent with the increased longevity and 15% lower mortality and cancer death rates seen in the seven western states with high natural background radiation averaging about 1mGy per year above that of the other states.[1-3,7]

The non-nuclear workers' death rates exactly matched those of the external non-shipyard matched control population. This demonstrates absence of the external healthy worker effect ascribed to adequate income, better health care, and the presence of health sufficient to allow maintenance of a reliable work schedule. There remains the question of an internal healthy worker effect resulting from the possible selection of more active individuals to be nuclear workers. The NW$_{>5}$ group with the greater exposure had a death rate from all causes of 0.76 the standardized mortality rate (SMR),

Table 11.1. Health Effects of Low Level Radiation in Shipyard Workers

Cause of Death	NW$_{>5}$	NW$_{<5}$	NNW
All Causes	2,797	1,168	4,453
SMR	0.76	0.81	1.00
(95% C.I.)	(0.73, 0.79)	(0.76, 0.86)	(0.97, 1.03)
Leukemia	21	4	29
SMR	0.91	0.42	0.97
(95% C.I.)	(0.56, 1.39)	(0.11, 1.07)	(0.65, 1.39)
LHC[a]	50	13	84
SMR	0.82	0.53	1.1
(95% C.I.)	(0.61, 1.08)	(0.28, 0.91)	(0.88, 1.37)
Mesothelioma	18	8	10
SMR	5.49	6.14	2.54
(95% C.I.)	(3.03, 8.08)	(2.48, 11.33)	(1.16, 4.43)
Lung Cancer	237	98	306
SMR	1.07	1.11	1.15
(95% C.I.)	(0.94, 1.21)	(0.90, 1.35)	(1.02, 1.29)

[a]Lymphatic and Hematopoietic Cancers.

16 standard deviations below that of the non-nuclear worker group (NNW). The $NW_{<5}$ with lesser exposure had 0.81 SMR, about 8SD below the NNW. While a possible internal healthy worker effect could contribute to the lowered SMR of nuclear workers, comparison of the $NW_{>5}$ group with the $NW_{<5}$ group demonstrates that the group with the greater dose had the lower SMR with even greater statistical power. This provides very strong evidence that low levels of ionizing radiation are without health hazard.

Leukemia and Mortality of Atomic Bomb Survivors Exposed to Low Level Radiation

A recent article by Shimizu, et al.[10] concerning the effects of low level radiation in atomic bomb survivors concluded that analysis of dose response "in the less-than-0.5Sv region fails to indicate the presence of hormesis." They did not observe any significant decrease in the relative risks (RR) of (a) leukemia, (b) all cancers except leukemia, (c) lung cancer, (d) thyroid cancer, or (e) noncancer mortality. This conclusion is in agreement with the data shown for the three cancer groups (b,c,d), but appears inconsistent with the data presented for the RR of the leukemia and noncancer mortality groups.

The upper half of Figure 11.1 shows the data for these two groups as analyzed by the authors with a variety of models. The discussion of leukemia states that though the RR is less than 1 for the three groups with doses less than 0.1Sv, since all had p > 0.10 they did not differ statistically from unity and thus, were within the range of random variation. In clear contradiction to least square fits, the quadratic model for < 0.5Sv was considered to better fit the data than the linear-quadratic model for < 0.5Sv that demonstrated a RR of 0.78 at 0.11Sv. The lower half of Figure 11.1 shows analysis of the data with models that provide a better fit. The five data points for leukemia are fitted by an empirical polynomial function. The RR for the 0.010 to 0.019, 0.020 to 0.049, and 0.050 to 0.099 Sv dose categories appear consistently related to one another, not varying randomly. The RR of 0.6 plotted at 0.075Sv is 1.5SD less than 1(p < 0.15). This study of atomic bomb survivors is in agreement with the decreased leukemia mortality seen in the nuclear shipyard worker study. In both studies the very low incidence of leukemia makes it difficult to obtain sufficient numbers for high statistical power.

Desired statistical power is present, however, for mortality rates. In the upper half of Figure 11.1 the RR data for noncancer mortality after low-level radiation are ignored and fitted with a threshold model derived from a prior study of survivors in the < 4.0Sv high-level dose range, assuming the threshold dose is 1.5. Though the mortality RR of 0.83 in the 0.200 to 0.499 Sv dose category is 3.2 SD below 1 (p = 0.001) and is the most statistically significant data point of the entire study, nevertheless, this highly significant decreased RR is rejected with the statement, "The RRs for the sub-

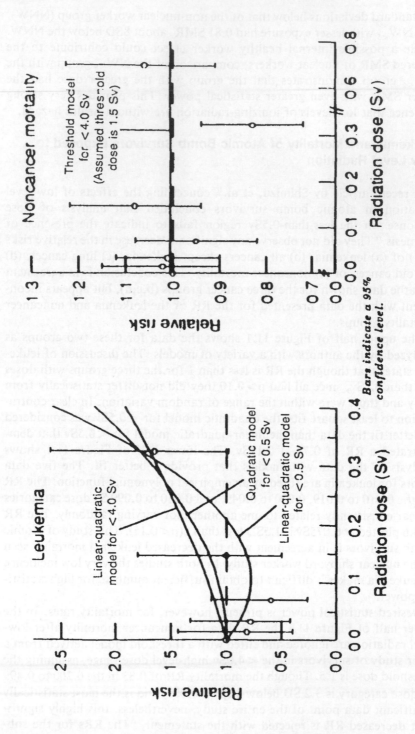

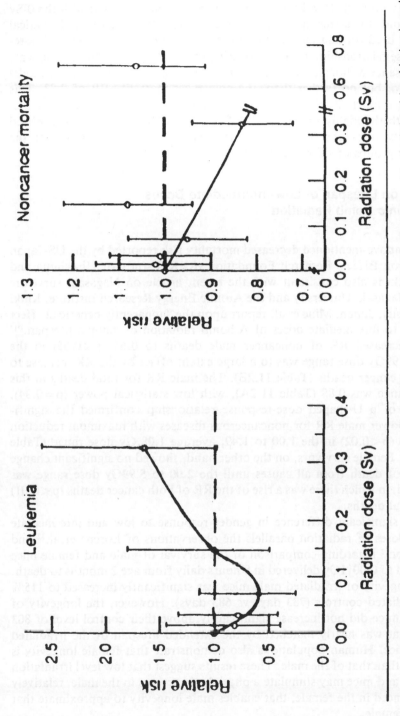

Figure 11.1. Dose-response analysis of atomic bomb survivors exposed to low-level radiation. The upper pair of relative risks of leukemia and noncancer mortality show the best fit models of the author's to their data. The lower pair of relative risks show the best fit models of the author of this review to their data.

groups within the low dose group (<0.5Sv) when compared with the 0-Sv group did not differ and were close to unity." If the only mathematical models used for analysis are those that a priori exclude a U-shaped dose-response relationship, it is not surprising that such analysis "fails to indicate the presence of hormesis." The lower half of Figure 11.1 fits a linear model down to, but no farther than, the noncancer mortality RR of 0.83. This decreased mortality risk associated with acute low-level radiation is consistent with the highly significant (-16SD and -8SD) decreased standardized mortality rates observed in prolonged very low level exposures of the nuclear groups of shipyard workers.[8]

Effect on Lifespan of Low-Intermediate Doses of Atomic Bomb Radiation

The above-mentioned decreased mortality risk reported by the US-Japan Radiation Effects Research Foundation (RERF) study of Hiroshima and Nagasaki is also consistent with the recent article on Nagasaki survivors from Nagasaki University and the Atomic Energy Research Institute, Kinki University, Japan. Mine et al. report upon the "apparently beneficial effect of low to intermediate doses of A-bomb radiation on human lifespan."[11] The decreased RR of noncancer male deaths to 0.65 (p<0.05) in the 0.50–0.99Gy dose range was to a large extent offset by the RR increase to 1.56 in cancer deaths (Table 11.2B). The male RR for total deaths in this dose range was 0.88 (Table 11.2A), with low statistical power (p = 0.34). Fitting of a U-shaped dose-response relationship confirmed the significantly lower male RR for noncancerous diseases with maximum reduction to 0.76 (p<0.02) in the 1.00 to 1.49, average 1.08,Gy dose range (Table 11.2C). Female survivors, on the other hand, showed no significant change in RR of death from all causes until the 2.00 to 5.99Gy dose range was reached, in which there was a rise of the RR of both cancer deaths (p<0.01) and total deaths.

This significant difference in gender response to low and intermediate acute doses of radiation parallels the observations of Lorenz et al.[12] and Congdon[13] regarding comparison of the survival of male and female mice exposed to 0.0011Gy delivered in 8 hours daily from age 2 months to death. The longevity of irradiated male mice was significantly increased to 115% of irradiated controls (783 days vs 683 days). However, the longevity of female mice did not increase significantly above their control level of 803 days that was nearly matched by the extended lifespan of the irradiated male mice. Human populations also demonstrate that female longevity is greater than that of the male. These results suggest that low level irradiation of men and mice may stimulate a physiologic process in the male, relatively unenhanced in the female, that enables male longevity to approximate that of the female.

Table 11.2.   Total Deaths and Relative Risks of Male and Female A-Bomb Survivors in Nagasaki During 1970-1988 Classified by T65 Dose

A. Initial numbers of subjects (1970), observed (O) and expected (E) numbers of total deaths and relative risk in 1970-1988 in Nagasaki among A-bomb survivors classified by T65D dose and sex.

T65D Dose (cGy)	Initial No. of Subjects		Total Deaths				Relative Risk (O:E)	
			Observed		Expected		All Causes	
	M	F	M	F	M	F	M	F
1–49	562	938	162	202	106·7	209	1·01	0·97
50–99	182	168	56	39	63·3	34·7	0·88	1·12
100–149	108	158	36	39	39·7	34·7	0·91	1·12
150–199	196	267	59	48	58·7	48	1·01	1·00
200–599	440	437	172	79	149·7	59·3	1·15	1·33

B. Observed (O) and expected (E) deaths in 1970-1988 in Nagasaki among A-bomb survivors classified by natural causes of death, sex and T65D dose.

Dose (cGy)	Number of Deaths from:			
	Non-Cancerous Diseases		Cancer	
	O	O:E	O	O:E
Males				
1–49	126	1·09	35	0·84
50–99	30	0·65	26	1·56
100–149	23	0·77	13	1·34
150–199	38	0·84	21	1·58
200–599	113	1·07	54	1·32
Females				
1–49	144	0·89	56	1·24
50–99	30	1·11	8	1·10
100–149	26	0·96	13	1·86
150–199	31	0·84	16	1·60
200–599	50	1·11	28	2·11

C. Calculated (L) values by the logistic function $p = 1/[1 + \exp\{-a - b_1(D - <D>) - B_2(D - <D>)^2 - cA\}]$ and observed (O) values for deaths from non-cancerous diseases in males in Nagasaki classified by T65D dose.

Dose (D) (cGy)	Number of Non-Cancer Deaths			
	Observed (O)	Corrected O:E	Calculated (L)	Corrected L:E
27 (1–49)	126	1·07	123	1·05
79 (50–99)	30	0·68	35·3	0·80
108 (100–149)	23	0·80	21·9	0·76
167 (150–199)	38	0·88	35·3	0·82
288 (200–599)	113	1·11	113·1	1·11

$<D> = 130$; $a = -6.14$ ($p < 0.01$); $b_1 = 0.29 \times 10^{-3}$ (p NS); $b_2 = 0.213 \times 10^{-4}$ ($p < 0.02$); $c = 0.115$ ($p < 0.01$).

NS = not significant.

Correlation of Lung Cancer Risk with Radon in Homes

The BEIR IV report [14] based upon a linear-no threshold extrapolation of the incidence of lung cancer in uranium mine workers exposed to high radon concentrations, predicts that the lifetime mortality risk of lung cancer is increased linearly by 10.8% per pCiL^{-1}. One pCiL^{-1} approximates the world average[15] and is equivalent to 0.2 working-level-month (WLM).[16] The American Cancer Society projects for the United States 170,000 new cases of lung cancer in 1993.[17] Accordingly, prior continued home exposure of the population to one additional pCiL^{-1} of radon would have produced 18,000 additional new cases of lung cancer in 1993. Five-year survival of treated lung cancer is only slightly more than 10%.[17] Relying upon the BEIR IV theoretical prediction, the Environmental Protection Agency (EPA) considers radon in the home to be the nation's leading health hazard.

However, there is no epidemiologic evidence to support the risks predicted by BEIR IV. To the contrary, epidemiologic studies in the United States,[18-20] Sweden,[21] Finland,[22] and China,[23] with increased radon concentrations up to 12 pCiL^{-1}, as well as in those areas below the average radon concentration of 1pCiL^{-1},[24-26] have all demonstrated a negative correlation of lung cancer with radon concentration. For a variety of reasons, these studies which contradict the linear-no threshold theory have been considered invalid by the National Academy of Sciences Committee on Biological Effects of Ionizing Radiation, National Council on Radiation Protection and Measurements, and the International Commission on Radiologic Protection. Criticisms have included poor statistical power, inadequate controls, and inadequate determination of the degree to which data have been altered by smoking and confounding factors such as numerous socioeconomic variables, geography, altitude, and climate. An extensive University of Pittsburgh National Survey of radon in homes was completed in 1992 that addresses these criticisms with excellent statistical power.

The University of Pittsburgh nationwide study based upon 272,000 measurements in the homes of 1217 counties was completed in 1992. This study and nine individual state studies were normalized to the EPA National Residential Radon Survey. The combined data set compiled from Pittsburgh, states, and EPA studies includes 1729 counties containing nearly 90% of the U.S. population. After deleting Arizona, California, and Florida, states with high retirement migration, and counties with incomplete data, 1601 counties remain included.[27] Figure 11.2 shows plots of mean age-adjusted lung cancer mortality rates (m) for white males (Figure 11.2a) and females (Figure 11.2c) vs mean radon levels (r) in homes of all counties within various ranges of r, along with the standard deviation of the mean, first and third quartiles, and the best linear fit to the data for individual counties, $m = 1a(1 + br)$. These mortality rates are corrected for smoking by use of Bureau of Census Population Surveys of smoking prevalence and BEIR IV risk estimates for smokers and nonsmokers, and are shown

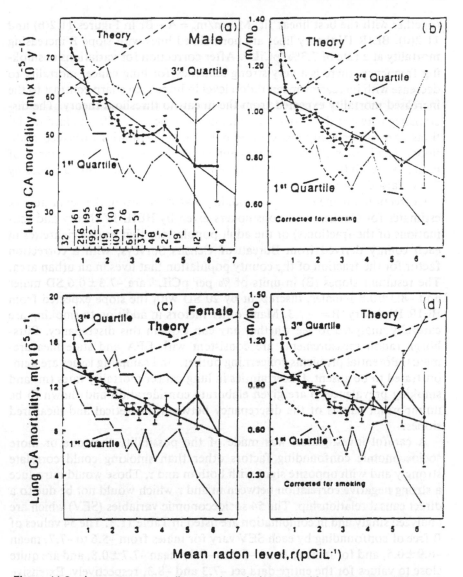

Mean radon level, r(pCiL⁻¹)

Figure 11.2.   Lung cancer mortality rates in male (11.2a) and female (11.2c) residents vs mean radon level for 160 counties. Data points shown are average of ordinates for all counties within the range of r-values shown on the baseline of Figure 11.2a, the number of counties within that range is also shown there. Error bars are standard deviation of the mean, and the first and third quartiles of the distribution are also shown. Figure 11.1b, 1d are m/m_0 vs r which incorporate the effects of smoking prevalence. Theory lines are arbitrarily normalized lines increasing at a rate of $7.3\%/r_0$.

together with the best linear fit, $M = m/m_o = A + Br$ in Figures 11.2(b) and 11.2(d). BEIR IV theory lines are normalized lines with slope B increasing mortality at a rate of $7.3\%/pCiL^{-1}$. After correction for variations in smoking frequency, there is a very strong tendency for lung cancer mortality to decrease with increasing mean radon level in homes, in sharp contrast to the increased mortality expected from the linear-no threshold theory. The discrepancy between theoretical and measured slopes is by 20 standard deviations. An earlier study based upon data for 965 counties furnished additional details of methodology and somewhat less steep negative slopes of m/m_o vs r, with the discrepancy between theoretical and measured slopes by 7 standard deviations.[28]

Correction for the effects of smoking was made using the separate risk estimates for smokers and nonsmokers given by BEIR IV theory and estimations of the fraction(s) of the adult populations that smoke cigarettes in each county derived from Bureau of Census Surveys, with a correction factor for the fraction of the county population that lives in an urban area. The resultant slopes (B) in units of $\%$ per $pCiL^{-1}$ are -7.3 ± 0.6 SD males and -8.3 ± 0.8 females, discrepant by 20 SD with the slope expected from BEIR IV theory, $B = +7.3$. Many other factors in addition to smoking are carefully analyzed to see whether any can explain this discrepancy. Pittsburgh radon measurements are consistent with EPA and state measurements. Potential problems concerning outliers and sampling issues are demonstrated to be absent. Uncertainties in lung cancer mortality rates (m) and smoking prevalence(S) are given elaborate consideration and shown to be unimportant causes of the discrepancy between theoretical and measured slopes.

A careful investigation was made of the possibility that one or more socioeconomic confounding factors other than smoking could correlate strongly and with opposite signs with both m and r. Those would introduce a strong negative correlation between m and r which would not be due to a direct causal relationship. The 54 socioeconomic variables (SEV) which are analyzed singly and in combination are listed in Table 11.3. The 54 values of B free of confounding by each SEV vary for males from -5.6 to -7.7, mean -6.9 ± 0.5, and for females from -5.4 to -9.1, mean -7.7 ± 0.8, and are quite close to values for the entire data set -7.3 and -8.3, respectively. Extensive statistical analysis of the possibility that some combination of SEV may act cooperatively to confound the m-r relationship concluded that the actual effect of confounding by combinations of SEV is to reduce the discrepancy between slopes by no more than 10%. Confounding by geography was also analyzed by considering the 34 states with at least 20 counties having known radon levels. The average of B-values is -6.1 for males and -7.2 for females; reductions in the discrepancy by 8.2% and 7.1%, respectively.

In addition to the 54 SEV and geography, also considered are the possible confounding physical features of altitude, average winter and summer temperatures, inches of precipitation, number of days per year with more than

Table 11.3.   Socioeconomic Variables Used in the Analysis for Radon Radiation Lung Carcinogenesis in the Low Dose, Low Dose Rate Region

Population Characteristics		Economics	
PT–	total population	EI–	$ per capita income
PD–	population/square mile	EH–	median household inc., $
PI–	% pop. increase 1980–86	EJ–	% persons below poverty level
PU–	% in urban areas	EV–	% fam below poverty level
PW–	% white	EU–	% unemployment
PS–	males/100 females	EW–	average salary, wage
PE–	% age > 64y	EP–	$ per cap personal income
PO–	% age > 74Y	EM–	% earnings from manufact.
PY–	% 5–17 years old	ER–	% earnings from retail trade
PN–	% born in state	ES–	% earnings from services
PH–	persons/household	EG–	% earnings from government
		EF–	% earnings from farming
Vital and Health Statistics		EA–	av. acres per farm
		EL–	% mfg. firms > 100 emplys.
VS–	births/1000 pop.	ED–	$/cap. sales–food stores
VC–	% births to mothers < 20y	EC–	$/cap. sales–clothing
VD–	deaths/1000 pop.	EX–	$/cap. sales–eating, drink
VI–	infant deaths/1000 births		
VM–	marriages/1000 pop.	**Government**	
VS–	divorces/1000 pop.	GF–	federal govt., $/cap
VP–	physicians/100,000 pop.	GL–	local govt., $/cap
VH–	hospital beds/100,000 pop.	GE–	% loc govt. expend.–educ.
		GH–	% loc govt. expend.–health
Social		GP–	% loc govt. expend.–police
		GW–	% loc govt. expend.–welf
SS–	social sec. benefit/1000 pop.	GR–	% loc govt. expend.–roads
SC–	crimes/100,000 pop.	GJ–	loc govt. emplmt/10,000 Pop.
SH–	% high school grad.	GV–	% vote for lead party, 1984
SU–	% college grad.		
SE–	$/cap for education		

Housing

HO–	% owner occupied
HA–	% with >1 automobile
HV–	median value ($)
HN–	% < 8 years old

0.01 inch precipitation, average wind speed, and percent of time with sunshine as compared with the maximum possible. Studies of these physical features concluded that none is an important confounding factor. The strong decrease in lung cancer mortality rates corrected for smoking frequency with increasing radon exposure is found in only the low altitude states or only the high altitude states; in only the warmest or only the coldest; in only the wettest or only the driest; etc. It is also found in only the states selected where the physical features are close to average. The BEIR IV theoretical prediction of lung cancer mortality from radon exposure corrected for smoking, $M = m/m_o = A + Br$, does not take into account two recognized r-S correlations: (1) urban houses have 25% lower radon levels than rural houses and urban people smoke more frequently, and (2) houses of smokers have 10% lower radon levels than houses of nonsmokers. An extensive statistical study of the effects of these r-S correlations leads to the conclusion that the BEIR IV prediction of B is reduced from $+7.3$ to $+6.9$, which contributes very little to decreasing the discrepancy with the large negative values of B, -7.3 and -8.3 obtained from the actual measured and reported data.

Linear-no threshold theories other than BEIR IV are considered which involve different treatments of smoking. Also considered is the "intensity of smoking." Analytic statistical study of these considerations lead to the conclusion that other theoretical predictions of B could reduce the discrepancy between 3% and 8%. The possibility that an unrecognized confounding factor could explain the discrepancy is recognized. However, the following properties are required of an unrecognized confounder that could resolve the discrepancy: (1) It must have a very strong correlation with lung cancer, comparable to that of cigarette smoking, but still be unrecognized; (2) It must have a very strong correlation of opposite sign with radon levels; (3) It must *not* be strongly correlated with any of the 54 socioeconomic variables (SEV); (4) It must be applicable in a wide variety of geographic areas and independent of altitude and climate. The first property alone requires of the unrecognized confounder that it must have increased by orders of magnitude since the beginning of this century, and have been much more important in males in the first half of the century, with effects on females rapidly catching up in recent years. The remaining properties impose additional requirements that are also most difficult to meet singly, while to satisfy the four simultaneously becomes incredible. These multiple restrictions upon an unknown confounder make it extremely improbable that one exists that would resolve the discrepancy.

These tests of the linear-no threshold theoretical prediction of lung cancer mortality induced by radon exposure, with the slope of the line determined by high dose exposures, demonstrate that the theory fails badly by gross overestimation of mortality in low dose, low dose rate range of radiation. A likely explanation is that stimulated biological defense mechanisms more than compensate for the radiation "insult" and are protective against cancer in a low dose, low dose rate range.

Breast Cancer in Women Exposed to Low Level Radiation

The Canadian study of fluoroscoped women includes 31,710 patients admitted to national sanitoriums between 1930 and 1952 and alive on January 1, 1950.[29] The results relate deaths from breast cancer between 1950 and 1980 that occurred 10 or more years after first exposure to fluoroscopic radiation. Fluoroscopic examination in Nova Scotia was performed AP (anterior-posterior), with the patient facing the fluoroscope. This position increased the breast dose to 50mGy per exposure compared to 2mGy per exposure in all the other provinces in which the examination was performed PA (posterior-anterior), with the patient's back against the fluoroscope. The standardized mortality rates from breast cancer for various dose ranges is shown in Table 11.4 with the high dose, high dose rate data of Nova Scotia separated from the low dose rate data of the other provinces.

Linear and linear-quadratic dose-response models were compared with respect to data fit. The authors concluded "that the most appropriate form of dose response relation is a simple linear one, with different slopes for Nova Scotia and the other provinces." On the basis of this linear model, Table 11.5 predicts the lifetime excess risk of death from breast cancer after a single exposure to 1cGy, an amount approximately three times the average annual background radiation.

The epidemiologic data listed in Table 11.4 and the associated fitted models were not presented graphically. The omitted graph is shown in Figure 11.3, together with an empirical polynomial function fitted to the data. The linear model for 2mGy exposures discards the data at 0.15Gy and

Table 11.4. Canadian Study of the Incidence of Breast Cancer Following Fluoroscopic Examinations

| Dose Gy | Standardized Rate Per 10⁸ Person Years | | |
	Nova Scotia	Other Provinces	All Provinces
0–0.09	455.6	585.8	578.6
	(131)	(288)	(301)
0.10–0.19		389.0	421.8
		(29)	(32)
0.20–0.29		497.8	560.7
		(24)	(26)
0.30–0.39	1709	630.5	650.8
	(11)	(17)	(18)
0.40–0.69		632.1	610.0
		(19)	(19)
0.70–0.99			1362
			(13)
1.00–2.99	2060		1382
	(14)		(17)
3.00–5.99	2811	873.1	2334
	(13)	(14)	(14)
6.00–10.00	7582		8000
	(8)		(9)
≥10.00	21.810		20.620
	(12)		(13)

ᵃThe number of deaths is shown in parentheses. The calculations exclude the values for 10 years after the first exposure and have been standardized according to age at first exposure (10 to 14, 15 to 24, 25 to 34, and ≥35 years) and time since first exposure (10 to 14, 15 to 24, 25 to 34, and ≥35 years) to the distribution for the entire cohort.

Table 11.5. Predicted Lifetime Excess Risk of Death from Breast Cancer per Million Women after a Single Exposure to 1 cGy

Age at Exposure Yr.	Additive-Risk Model	Relative-Risk Model
10	125	108
20	95	89
30	67	55
40	42	27

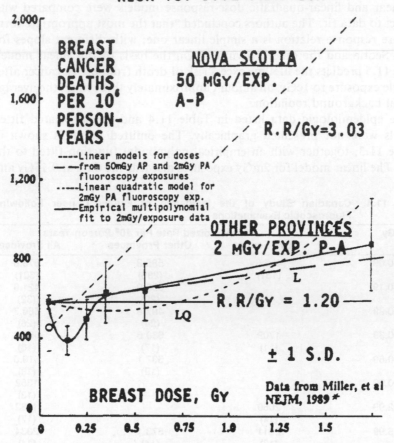

CANADIAN TB FLUOROSCOPY STUDY
RISK VS. DOSE FRACTION

BREAST
CANCER
DEATHS
PER 10⁶
PERSON-
YEARS

NOVA SCOTIA
50 mGy/EXP.
A-P

R.R/Gy=3.03

----Linear models for doses
— —from 50mGy AP and 2mGy PA
fluoroscopy exposures
·····Linear quadratic model for
2mGy PA fluoroscopy exp.
——Empirical multipolynomial
fit to 2mGy/exposure data

OTHER PROVINCES
2 mGy/EXP. P-A

L

R.R/Gy = 1.20 ——

LQ

± 1 S.D.

BREAST DOSE, Gy   Data from Miller, et al
NEJM, 1989 *

Figure 11.3. A graphic plot of the authors' data shown in Table 11.4. No graph was presented in their publication. The figure includes their "best fit" linear models, their linear quadratic model, and the best fit model of the author of this review to their data, an empirical polynomial function.

at 0.25Gy, the data with the best confidence limits. Compared to the controls receiving 0 to 0.09Gy, 0.15 Gy and 0.25 Gy demonstrate relative risks (RR) of 0.66 (p < .01) 0.85 (p < .38), respectively. While the RR of 0.85 is not statistically significant, it is consistent with the significant RR of 0.66 and the zero equivalent point of 0.31Gy indicated by the fitted polynomial function. For exposures above the zero equivalent point, the RR becomes positive after being negative in the range of 0 to 0.31 Gy. The decreased RR of breast cancer produced by low dose, low dose rate radiation were rejected a priori by the choice of mathematical models that extrapolate the dose-risk relation from high dose exposures to low dose exposures. The risks associated with low dose exposures cannot be measured, the authors state, "because the expected small excess of breast cancers would be obscured by the much higher background rate of breast cancer." Consequently, the unexpected was rejected since the possibility of a measurable decreased risk associated with low exposures appeared to be inconceivable. The highly significant decreased RR of 0.66 at 0.15Gy and the RR of 0.85 at 0.25Gy, both with the highest confidence limits of the entire study, are not shown graphically, not even discussed. Instead, the linear model for 0.002Gy exposures is used in Table 11.5 to predict the lifetime excess risk of death from breast cancer to be approximately 60 per million women after a single exposure to 1cGy at the age of 30. Nine hundred excess deaths from breast cancer are predicted theoretically from the exposure of one million women to 0.15Gy. However, the quantified low dose data predicts with better than 99% confidence limits that instead of causing 900 deaths, a dose of 0.15Gy would prevent 10,000 deaths in these million women.

Significant positive health effects associated with low level radiation have been demonstrated in a review of five epidemiologic studies: decreased mortality of nuclear shipyard workers, decreased noncancer mortality of atomic bomb survivors in both Hiroshima and Nagasaki and Nagasaki alone, decreased lung cancer mortality associated with increased radon exposure of the U.S. population, and decreased breast cancer mortality of women in Canada after having received multiple fluoroscopic examinations. The tendency to neglect or reject data that contradicts the linear-no threshold theory of radiation carcinogenesis is supported by confidence that chromosome aberration and gene mutation can be produced by a single particle of ionizing radiation and so initiate a malignancy. The number of such interactions with cell nuclei is both logically and demonstrably proportional to the dose. However, no consideration is given to biological defense mechanisms that could be stimulated further by low level increments of radiation above the background level. Such stimulated defense mechanisms could also decrease carcinogenesis by chemical and other non-ionizing agents as well as high level ionizing radiation. Multiple defense mechanisms at molecular, cellular, organ, and systemic levels involving enzymatic, hormonal, immunologic, and stress protein interactions are currently being demonstrated and confirmed by numerous investigators.[30-32] Recently a human radiation repair gene has been cloned and transfected into a mutant Chinese hamster with sensitivity to both ionizing radiation and certain alkylating agents resulting from defective repair of DNA strand breaks. These transfected mutants demonstrate overexpression of the human DNA repair minigene with repair capacity increased above that of the wild-type Chinese hamsters.[33]

Mounting reproducible evidence of the operation of various defense mechanisms and their stimulation by low dose ionizing radiation will provide further details of how biological defense mechanisms, nonoperative at high doses, are stimulated and enhanced by low level radiation damage so as to overcorrect and predominate. These investigations have clarified why the negative health effects observed at high levels of radiation that effectively overwhelm these defense mechanisms cannot be extrapolated to the low levels in which these stimulated defense mechanisms predominate with decreased cancer induction, decreased mortality, and other observed positive health effects.

REFERENCES

1. Craig, L., and H. Seidman. "Leukemia and Lymphoma Mortality in Relation to Cosmic Radiation," *Blood* 17:319 (1961).
2. Frigerio, N.A., and R.S. Stowe. "Carcinogenic and Genetic Hazard from Background Radiation," *Biological and Environmental Effects of Low Level Radiation*, Vol. 2: 285, International Atomic Energy Agency, Vienna (1976).
3. Webster, E.W. "The Effects of Low Doses of Ionizing Radiation," *J. Tenn. Med. Assoc.* 76:499 (1983).
4. Nambi, K.S.V., and S.D. Soman, "Environmental Radiation and Cancer in India," *Health Phys.* 52:653 (1987).
5. Nambi, K.S.V., and S.D. Soman. "Further Observations on Environmental Radiation and Cancer in India," *Health Phys.* 59:339 (1990).
6. Zahi, S., X. Lin, T. Pan, W. He, R. Feng, M. Chen, S. Li, L.R. Chen, and H. Yie. "Report of Survey on Mortality from Malignant Tumors in High Background Area of Guangdong," *J. Radiat. Res. (Japan)* 22:48 (1982).
7. Cohen, J.J. "Natural Background as an Indicator of Radiation Induced Cancer," *Proc. 5th Int. Ratiat. Protect. Assoc*, Jerusalem (1980).
8. Cameron, J. "The Good News about Low Level Radiation Exposure: Health Effects of Low Level Radiation in Shipyard Workers," *Health Phys. Soc. Newsletter* 20:9 (1992).
9. National Academy of Sciences Committee on the Biological Effects of Ionizing Radiation (BEIR V), "Health Effects of Exposure to Low Levels of Ionizing Radiation," National Academy Press, Washington, DC (1990).
10. Shimizu, Y., H. Kato, W.J. Schull, and K. Mabuchi. "Dose-Response Analysis of Atomic Bomb Survivors Exposed to Low-Level Radiation," *Health Physics Society Newsletter* 21:1 (1992); *RERF Update* 4,3:3 (1992).
11. Mine, M., Y. Okumura, M. Ichimara, T. Nakamura, and S. Kondo. "Apparently Beneficial Effect of Low to Intermediate Doses of A-Bomb Radiation on Human Lifespan," *Int. J. Radiat. Biol.* 58:1035 (1990).
12. Lorenz, E., J.W. Hollcroft, E. Miller, C.C. Congdon, and R. Schweisthal. "Longterm Effects of Acute and Chronic Irradiation in Mice. I Survival and Tumor Incidence Following Chronic Irradiation of 0.11r per Day," *J. Natl. Cancer Inst.* 15:1409 (1955).
13. Congdon, C.C. "A Review of Certain Low Level Ionizing Radiation Studies in Mice and Guinea Pigs," *Health Physics* 52:593 (1987).
14. National Academy of Sciences Committee on the Biological Effects of Ionizing Radiation (BEIR IV), "Health Risks of Radon and Other Internally Deposited Alpha Emitters," National Academy Press, Washington, DC (1988).

15. United Nations Scientific Commission on Effects of Atomic Radiation, "Ionizing Radiation, Sources and Biological Effects," United Nations, New York (1982).
16. NCRP84 National Council on Radiation Protection and Measurements, "Evaluation of Occupational and Environmental Exposures to Radon and Radon Daughters in the United States," NCRP Report No. 78, NCRP, Bethesda, MD (1984).
17. American Cancer Society, Boring, C.C., T.S. Squires, and T. Tong. "Cancer Statistics, 1993," *CA Cancer J. Clin.* 43:7 (1993).
18. Cohen, B.L. "Tests of the Linear, No-Threshold Dose-Response Relationship for High-Let Radiation," *Health Phys.* 52:629 (1987).
19. Cohen, B.L. "A National Survey of Radon in Homes and Correlating Factors," *Health Phys.* 51:2 (1986).
20. Cohen, B.L. "Expected Indoor ^{222}Rn Levels in Counties with Very High and Very Low Lung Cancer Rates," *Health Phys.* 57:897 (1989).
21. Swedjemark, G.A. and L. Mjones. "Radon and Radon Daughter Concentration in Swedish Homes," *Rad. Protection Dosimetry* 7:341 (1984).
22. Castren, O., K. Winquist, I. Makaleinen, and A. Voutilainen. "Radon Measurements in Finnish Houses," *Rad. Protection Dosimetry* 7:333 (1984).
23. Hofman, W., R. Katz, and Z. Chunxiang. "Lung Cancer Risk at Low Dose of Alpha Particles," *Health Phys.* 51:457 (1986).
24. George, A.C. and A.J. Breslin. "The Distribution of Ambient Radon and Radon Daughters in Residential Building in the New Jersey-New York Area," in: The Natural Radiation Environment III, T.F. Gesell, and W.M. Lowder, Eds., U.S. Dept. of Energy, Washington, DC (1980).
25. Nero, A.V., M.L. Boegel, C.D. Hollowell, J.G. Ingersoll, and W.W. Nazaroff. "Radon Concentrations and Infiltration Rates Measured in Conventional and Energy-Efficient Houses," *Health Phys.* 45:401 (1983).
26. Wrixon, A., L. Brown, K.D. Cliff, C.M.H. Driscoll, B.M.R. Green, and J.C.H. Miles. "Indoor Radiation Survey in UK," *Rad. Protection Dosimetry* 7:321 (1984).
27. Cohen, B.L. "Test of the Linear-No Threshold Theory of Radiation Carcinogenesis in the Low Dose, Low Dose Rate Region," *Health Phys.* in press (1994).
28. Cohen, B.L. and G.A. Colditz. "Tests of the Linear-No Threshold Theory for Lung Cancer Induced by Exposure to Radon," *Environmental Res.*, 64:65 (1994).
29. Miller, A.B., G.R. Howe, G.J. Sherman, J.P. Lindsay, M.J. Yaffe, P.J. Dinner, H.A. Risch, and D.L. Preston. "Mortality from Breast Cancer after Irradiation during Fluoroscopic Examinations in Patients Being Treated for Tuberculosis," *New England Journal of Med.* 321:1285 (1989).
30. Calabrese, E.J., Ed. *Biological Effects of Low Level Exposures to Chemicals and Radiation*, Lewis Publishers, Chelsea, MI (1992).
31. Luckey, T.D. *Radiation Hormesis*, CRC Press, Boca Raton, FL (1991).
32. Sugahara, T., L.A. Sagan, and T. Aoyama, Eds. *Low Dose Irradiation and Biological Defense Mechanisms*, Elsevier Science Publishers, Amsterdam (1992).
33. Caldecott, K.W., J.D. Tucker, and L.H. Thompson. "Overcorrection of the CHO EM9 DNA Repair Mutant by Overexpression of a Human XRCC1 Minigene," *Nucl. Acid Res.* 20:4575 (1992).

CHAPTER 12

Test of the Linear-No Threshold Theory of Radiation Carcinogenesis

Bernard L. Cohen, Physics Department,
University of Pittsburgh, Pittsburgh, Pennsylvania

We recently completed a compilation of radon measurements from available sources which gives the average radon level, r, in homes for 1730 counties, well over half of all U.S. counties, and comprising about 90% of the total U.S. population. Plots of age-adjusted lung cancer mortality rates, m, vs these r are shown in Figure 12.1a, c where, rather than showing individual points for each county, we have grouped them into intervals of r (shown on the baseline along with the number of counties in each group) and we plot the mean value of m for each group, its standard deviation, and the first and third quartiles of the distribution. We see, in Figure 12.1a, c, a clear tendency for m to *de*crease with increasing r, in sharp contrast to the *in*crease expected from the fact that radon can cause lung cancer, shown by the line labeled "theory."

One obvious problem is migration: people do not spend their whole life and receive all of their radon exposure in their county of residence at time of death. However, it is easy to correct the theoretical prediction for this, and the "theory" lines in Figure 12.1 have been corrected. As part of this correction, data for Florida, California, and Arizona, where many people move after retirement, have been deleted, reducing the number of counties to 1601. (This deletion does not affect results.)

A more serious problem is that Figure 12.1 is what epidemiologists call an "ecological study." Epidemiologists normally study the relationship between mortality risks to individuals, m', vs their personal exposure, r', whereas an ecological study like ours deals with the relationship between the average risk to groups of individuals (populations of counties) and their average exposure. It is well known to epidemiologists that, in general, the average dose does *not* determine the average risk, and to assume otherwise is called "the ecological fallacy." However, it is easy to show[1] that, in testing a linear-no threshold theory, "the ecological fallacy" does not apply; in that theory,

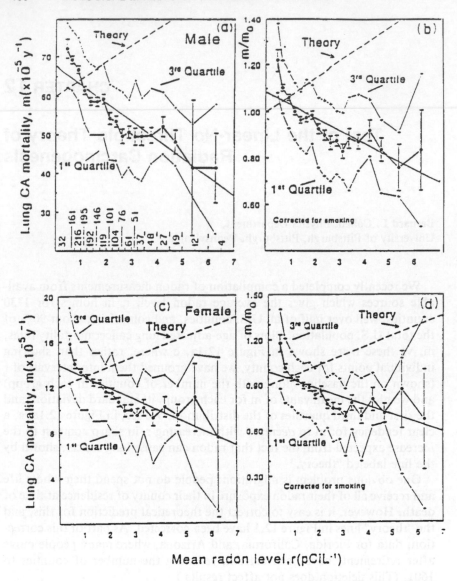

Figure 12.1. Lung cancer mortality rates vs mean radon level in homes for 1601 U.S. counties.

the average dose *does* determine the average risk. This is widely recognized from the fact that "person-rem" determines the number of deaths. Dividing person-rem by population gives average dose, and dividing number of deaths by population gives mortality rate.

Because of the "ecological fallacy," epidemiology textbooks often state that an ecological study cannot determine a causal relationship between risk and exposure. That may be true, but it is irrelevant here because the purpose of our study is *not* to determine a causal relationship; it is rather to test the linear-no threshold dependence of m on r.

Apart from "the ecological fallacy," other potential problems with ecological studies have been pointed out by Morgenstern, Greenland, and Robins,[3,4,5] but these have been shown not to be applicable to our work.[2,6,7]

The most obvious potential explanation for Figure 12.1 is that there is a strong negative correlation between the percentage of adult population that smokes, S, and radon exposure, r; i.e., that counties with low r tend strongly to have high S, and vice versa. This effect is most easily handled by use of the BEIR-IV theory,[8] which can be shown to give

$$m' = a(1 + br') \tag{1}$$

where m' is the lung cancer mortality risk to an individual, r' is that individual's radon exposure, and a and b are constants with a given separately for smokers and nonsmokers (a_s, a_n) and for males and females. If we sum over all individuals in a county and divide by the population, Eq. 1 reduces to

$$m = [Sa_s + (1 - S)a_n] (1 + br). \tag{2}$$

Applying our correction for migration and inserting numerical values for a_s and a_n then leads to[9]

$$m/m_o = 1 + Br \tag{3}$$

where

$$m_o = 9 + 0.99S \quad \text{for males}$$
$$m_o = 3.7 + 0.32S \quad \text{for females} \tag{4}$$
$$B = + 7.3$$

with B in units of percent per pCi/L of average radon level, and m_o in units of deaths/year-100,000. In Eq. 3, m/m_o may be thought of as the lung cancer mortality rate corrected for smoking prevalence.

Problems in determining S will be discussed below. Using our best values to calculate m_o from Eq. 4 for each county leads to results shown in Figure 12.1b, d. We see that correcting for smoking does little to improve the unexpected behavior. Fitting the data to

$$m/m_o = A + Br \tag{5}$$

to determine A and B gives $B = -7.3 \pm 0.6$ for males and $B = -8.3 \pm 0.8$ for females, as compared with the Eq. 4 theory prediction $B = +7.3$, a discrepancy of about 20 standard deviations. We refer to this as "our discrepancy," and the remainder of this chapter deals with our attempts to explain it, each section treating a different approach.

Uncertainties in Radon Data

Our radon data derives from three independent sources, our own measurements, EPA measurements, and studies by agencies in various individual states. Various checks for consistency among these three sources give satisfactory results.[2] Data from each of these three sources alone gives results for B very similar to those from our combined data set. We conclude that uncertainties in our r-values are not responsible for any significant part of our discrepancy. In fact, the simplest correction for these uncertainties would *increase* our discrepancy by about 8%.

Outliers and Sampling Issues

The effects of outlying points in our analyses of data on m/m_o vs r was investigated by using five of the most popular statistical tests to discard either 10 or 20 outliers. In all cases, for both males and females, this *increased* our discrepancy. Outliers were not discarded.

Ten different random samples each of 200, 400, and 800 of our 1601 counties were analyzed independently. In all cases, results for B were quite similar to those for our entire data set, B = -7.3 for males and -8.3 for females. For example, for our ten random sets of 200 counties, all B values were between -5.0 and -8.5 for males, and between -4.8 and -12.7 for females. Our study might therefore be considered equivalent to eight independent studies, each giving roughly the same discrepancy with theory.

One might wonder how unexpected it is to find such a strong and statistically robust correlation between m and r as we find for lung cancer in Figure 12.1. To investigate this, we studied the regression of m on r for the 33 principal cancer types. The number of standard deviations by which the slope B differs from zero was 2.7 times larger for lung cancer than for any other type, and with just two exceptions it was at least 4 times larger. Double regressions on r and S gave similar results; as expected, the m-S correlation is very large and positive for lung cancer, and the m-r correlation is large (two-thirds as large as m-S). The only unexpected result was that the m-r correlation is negative rather than positive. We conclude that the strong observed correlation between m and r for lung cancer is quite unique and remarkable.

Uncertainties in Smoking Prevalence, S

Our S values were derived from a 1985 survey[10] of smoking prevalence in states, S', corrected for variations with time in national smoking prevalence[11] under the assumption that the ratio of S' for various states did not vary with time. It was then assumed that S values for the counties within a state are due only to urban-rural differences. That is, we take S = S'(1 + kPU)/(1 + kPU'), where PU is the percent of the population that lives in urban areas for the county, PU' is the same quantity for the state as a whole, and k is a constant determined from regressions of m on PU (k was found to be similar for all geographic regions).

An alternative method for determining S' values for states was by use of cigarette sales tax collections,[12] which are available for every year. This has the advantage of giving data for the relevant time periods and also reflects the number of cigarettes smoked rather than just the number of smokers, although it also has some recognized disadvantages. When these values of S' were used, our discrepancy was *increased*. They were not used further.

As an approach to getting direct data on S for counties in the relevant time period with due consideration for intensity of smoking (e.g., inhalation, cigarettes per day), we developed a smoking variable S derived from lung cancer mortality data. We utilized socioeconomic variables (SEV) listed in Table 12.1 plus S' to predict m-values in a manner independent of radon levels, r. We stratified on r into six separate groups of counties, and for each group independently, studied multiple regressions of m on SEV. We were able to derive a linear combination of S' plus five SEV with coefficients independent of r, which predict m-values about as well as they can be predicted from SEV. When S values derived from this process are used to calculate m_o from Eq. 4, and these are then used to fit the data to Eq. 5, B values are changed from −7.3 to −6.0 for males, and from −8.3 to −6.3 for females. Since this represents only a modest reduction in our discrepancy, and since it is questionable to use S-values derived from m-values to predict m-values, these S-values were not used in our other studies, but this exercise indicates that the obvious problems in our derivation of S-values are not the cause of our discrepancy.

As an entirely different approach to evaluating effects of uncertain S-values, we then set out to determine how strong a negative r-S correlation would be needed to explain our discrepancy. We re-assigned S-values for our 1601 counties in perfect reverse order of their r-values, and used these S-values in our analysis. This "perfect" negative r-S correlation reduced our B-values essentially to zero (+0.7 for males, −0.3 for females), only cutting our discrepancy in half. The problem is that our distribution of S-values is rather narrow — for males, mean −51.7, SD −6.9, min/max −25/70. If we arbitrarily double the width of this distribution by doubling the difference from the mean for each county to give mean −51.7, SD −13.8, min/max −0/88, we are able to eliminate our discrepancy by reassigning S-values in a manner that gives the coefficient of correlation (CORR) between S and r to be −0.90.

We then consider the question of how strong an r-S correlation is credible. Since any such correlation must arise from confounding by socioeconomic variables, we studied correlations of our 54 SEV (Table 12.1) with r. The largest |CORR-r| for any of our SEV is 0.37, the second largest is 0.30, and for 49 of our 54 SEV, |CORR-r| is less than 0.23. For the S-values we are using, CORR-r is −0.28 for males and −0.19 for females. It therefore

Table 12.1. Socioeconomic Variables (SEV) Used in the Analysis

Population Characteristics		Economics	
PT–	total population	EI–	$ per capita income
PD–	population/square mile	EH–	median household inc., $
PI–	% pop. increase 1980–86	EJ–	% persons below poverty level
PU–	% in urban areas	EV–	% fam below poverty level
PW–	% white	EU–	% unemployment
PS–	males/100 females	EW–	average salary, wage
PE–	% age > 64y	EP–	$ per cap personal income
PO–	% age > 74Y	EM–	% earnings from manufact.
PY–	% 5–17 years old	ER–	% earnings from retail trade
PN–	% born in state	ES–	% earnings from services
PH–	persons/household	EG–	% earnings from government
		EF–	% earnings from farming
Vital and Health Statistics		EA–	av. acres per farm
		EL–	% mfg. firms > 100 emplys.
VS–	births/1000 pop.	ED–	$/cap. sales–food stores
VC–	% births to mothers < 20y	EC–	$/cap. sales–clothing
VD–	deaths/1000 pop.	EX–	$/cap. sales–eating, drink
VI–	infant deaths/1000 births		
VM–	marriages/1000 pop.	**Government**	
VS–	divorces/1000 pop.		
VP–	physicians/100,000 pop.	GF–	federal govt., $/cap
VH–	hospital beds/100,000 pop.	GL–	local govt., $/cap
		GE–	% loc govt. expend.–educ.
Social		GH–	% loc govt. expend.–health
		GP–	% loc govt. expend.–police
SS–	social sec. benefit/1000 pop.	GW–	% loc govt. expend.–welfare
SC–	crimes/100,000 pop.	GR–	% loc govt. expend.–roads
SH–	% high school grad.	GJ–	loc govt. emplmt/10,000 pop.
SU–	% college grad.	GV–	% vote for lead party, 1984
SE–	$/cap for education		
Housing			
HO–	% owner occupied		
HA–	% with >1 automobile		
HV–	median value ($)	NP–num of measurements–PITT	
HN–	% < 8 years old	NE–num of measurements–EPA	

seems incredible that the true r-S correlation can be of the magnitude necessary to explain our discrepancy, even if coupled with a large error in the width of our distribution of S-values. We conclude that uncertainties in S-values are not a major cause of our discrepancy.

Confounding by SEV and Factors that Correlate with Them

If a particular socioeconomic variable (SEV) is an important confounding factor (CF), stratifying our data on it into subsets and analyzing each subset separately would greatly reduce the problem as all counties in a given subset would have approximately the same value of that SEV. The average of the B-values obtained from the various subsets would then give a value of B free from the effects of confounding.

The data were stratified into five quintiles of 1601/5 = 320 counties on the basis of each of our 54 SEV in turn. This gave 540 subsets (including both sexes), and for all 540 of them, B was found to be negative. Thus, the negative slopes in Figure 12.1b, d are found if we consider only the most urban countries, or if we consider only the most rural; if we consider only the richest, or only the poorest; if we consider only those with the best medical care, or only those with the poorest medical care, etc., for our 54 SEV. They are also found if we consider any of the strata in between.

Following up on our method of averaging B-values over the five quintiles to obtain B-values free of confounding gives, for our 54 SEV, results ranging between –5.6 and –7.7 for males, and between –5.4 and –9.1 for females, reasonably close to our values for the entire data set, –7.3 and –8.3. We conclude that confounding by any one of our SEV can do little to explain our discrepancy.

This also excludes factors that correlate strongly with SEV as potential CF. For example, air pollution correlates strongly with several of our SEV (e.g., population) and therefore cannot be an important CF.

Confounding by Combinations of SEV

This still leaves open the possibility that some combination of SEV can explain our discrepancy. The best way to investigate this is through multiple regression analysis, fitting our data to

$$m/m_o = A + Br + c_1X_1 + c_2X_2 + \ldots c_{54}X_{54}$$ (6)

where $X_1 \ldots X_{54}$ are our 54 socioeconomic variables and A, B, $c_1 \ldots c_{54}$ are constants used to fit the data. With 1601 data points, there is no difficulty in deriving statistically robust estimates of these 56 constants. The results are B = –3.1 ± 0.6 for males, and B = –3.5 ± 0.9 for females, reducing our discrepancy by 29% and 31%, respectively.

However, the statistics community generally takes a dim view of using multiple regression on many variables to quantify the causal relationship of one particular variable. In our case, the strong negative correlation between m and r would cause any variable strongly correlated with m to have a correlation of opposite sign with r. In fitting Eq. 6, its term will therefore drain away some of the strength of the Br term, reducing the value of B.

As a study of this effect, Figure 12.2 shows a plot of CORR-m vs CORR-r for each of our SEV. We see there that every SEV with a large |CORR-r| has a large CORR-m of opposite sign, and vice versa. This could be a very remarkable coincidence, but it is much more credible that it is the result of the effect we are studying. This implies that the reduction in our discrepancy in going from simple to multiple regression is largely artificial, and the true values of B are close to –7.3 for males and –8.3 for females.

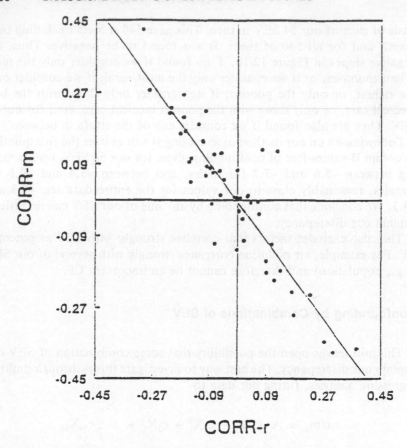

Figure 12.2. CORR-m vs CORR-r for socioeconomic variables listed in Table 12.1.

Confounding by Geography

It is well known that radon levels correlate strongly with geography,[1,13] which suggests that it be considered as a CF. We treat it by our stratification method.

The U.S. Bureau of Census divides the nation into four regions, each consisting of two or three divisions. Stratifying by regions and averaging B-values over the four strata gives B = -6.1 for males and -8.0 for females, reasonably close to our values without stratification, -7.3 and -8.3. However, stratifying on the 9 divisions gives an average B of -4.4 for males and -6.6 for females, a substantial reduction in our discrepancy. This suggests that finer stratification on geography may help explain our discrepancy.

The finest stratification readily available is by individual states. There are 34 states in which we have data on at least 20 counties. The average B-value from separate analysis of each of these is -6.1 for males and -7.2 for females. These reduce our discrepancy by 8% and 7%, respectively. We conclude that confounding by geography does little to reduce our discrepancy.

Confounding by Altitude and Weather

Rather different types of potential confounding factors are barometric pressure (determined by altitude) and weather. Data on these are available only by states.

If we treat data on states analogously to how we have been treating it for counties, we have only 46 data points instead of 1601, but an analogous analysis can be done. This gives B = -13.0 ± 2.3 for males and B = -14.4 ± 2.7 for females, as opposed to B = + 8.3 predicted by the theory, a very statistically robust discrepancy.

As potential CF we consider altitude (meters above sea level), average winter temperature, average summer temperature, millimeters of annual precipitation, days/year with measurable precipitation, average wind speed, and percent of time with sunshine. We stratify the data on the basis of each of these in turn into three subsets of 15-16 states and analyze each subset to determine B. This gives a total of 42 analyses for both sexes, and all 42 B-values are found to be negative. Averaging over the three strata gives B-values ranging for our seven variables from -9.0 to -15.5 for males and from -11.8 to -15.6 for females. In no case are the deviations from values without stratification for a given variable in the same direction for males and females, and in no case is the average deviation for the two sexes more than 0.6 SD. Large negative B-values are found if we consider only low altitude states or if we consider only high altitude states; if we consider only warm states, or only cool states; if we consider only wet states, or only dry states; etc. They are also found if we consider only states with average values of these properties. These properties cannot, therefore, be the cause of our discrepancy.

Effects of Recognized r-S Correlations

In our extensive studies of correlations with radon levels vs. house characteristics, locations, and socioeconomic factors,[13] we encountered two situations which would lead to r-S correlations:

1. Urban houses average 25% lower radon levels than rural houses, and urban people smoke 20% more frequently, judging from urban-rural differences in lung cancer rates.
2. Houses of smokers have 10% lower average radon levels than houses of nonsmokers.

A detailed calculation of the effects of these r-S correlations found that (1) changes the slope of an m vs r regression by 18%, but the effect is almost completely compensated by our correction for smoking, changing the slope, B, of an m/m_o vs r regression by less than 1%. The smoking correction does not compensate (2), but it changes the slope B by only 5%. Items (1) and (2) were found to add linearly in their effect on B.

These recognized r-S correlations, therefore, change B by only 6% and thus reduce our discrepancy by only about 3%. It seems most unlikely that there are unrecognized r-S correlations that are over an order of magnitude larger than these, as would be necessary to explain our discrepancy.

Dependence on BEIR-IV Theory

All calculations to this point, including our correction for smoking, have been carried out using the BEIR-IV theory;[8] however, we have shown that our discrepancy would be about equally large for any other m-r-S relationship based on data from the miners. The principal differences among competing theories are in their treatment of smoking, but since r-S correlations are not very strong, these differences have little effect on the results.

Unrecognized Confounding Factors

It is logically possible that there is some unrecognized confounding factor (UCF) which is causing our discrepancy. Of course a UCF could invalidate *any* epidemiological study, and few, if any, epidemiological studies have included as thorough investigation as ours of confounding factors.

Let us consider the properties of a UCF necessary to explain our discrepancy:

- it must have a very strong correlation with lung cancer, at least comparable to that of smoking, but still be unrecognized as such
- it must have a very strong correlation of opposite sign with radon levels
- it must not be strongly correlated with any of our socioeconomic variables, or with pressure, temperature, or other weather variables
- it must be operative in the great majority of geographical areas.

Requirement (1) means that in addition to causing lung cancer and being unrecognized as such, it must have increased in importance by orders of magnitude since the early part of this century; it must have affected males much more than females until mid-century, with females closing the gap in recent years; it must be an order of magnitude more important in smokers than in nonsmokers, etc. Requirement (2) is also difficult since correlations between radon and other factors have been studied extensively and are nearly all rather weak; also, factors affecting radon levels are well understood. Requirements (3) and (4) impose additional severe restrictions, taking away nearly all options that one would ordinarily consider. We therefore judge the existence of a UCF fulfilling all of these requirements to be essentially incredible, although we are always open to suggestions.

Conclusions

We have explored every explanation for our discrepancy that we can think of or that has been suggested to us. By far the most credible explanation, in our view, is that the linear-no threshold theory fails very badly in the low dose, low dose rate region where it has never been previously tested, grossly overestimating the cancer risk.

REFERENCES

1. Cohen, B.L. *Int. J. Epidemiol.* 19:680–684 (1990).
2. Cohen, B.L. *Crit. Rev. Environ. Control.* 22:243–364 (1992).
3. Morgenstern, H. *Am. J. Public Health* 72:1336–1344 (1983).
4. Greenland, S. and H. Morgenstern, *Int. J. of Epidemiol.* 18:269–274 (1989).
5. Greenland, S. and J. Robins, *Am. J. Epidemiol.* (in press)
6. Cohen, B.L. *Int. J. Epidemiol.* 21:422–424 (1992).
7. Cohen, B.L. *Am. J. Epidemiol.* (in press)
8. BEIR (National Acad. of Sciences Com. on Biological Effects of Ionizing Radiation). Health Risks of Radon. . . . National Academy Press, 1988 (BEIR-IV).
9. Cohen, B.L. and G.A. Colditz, *Environmental Res.* 64:65–89 (1994).
10. U.S. Public Health Service, Smoking and Health: A National Status Report. DHHS Publication 87-8396, 1990.
11. U.S. Public Health Service, Morbidity and Mortality Weekly Reports 36, 581–584 (1987).
12. Tobacco Institute, The Tax Burden on Tobacco, 1988.
13. Cohen, B.L. *Health Physics* 60:631–642 (1991).

CHAPTER 13

Bayesian Estimation of Incidence Change-Points in Low Dose Experimental and Epidemiologic Studies

Peter G. Groer, Department of Nuclear Engineering,
University of Tennessee-Knoxville, Knoxville, Tennessee

Bruce A. Carnes, Biological and Medical Research Division,
Argonne National Laboratory, Argonne, Illinois

INTRODUCTION

In 1953 I realized that the straight line leads to the downfall of mankind. But the straight line has become an absolute tyranny. The straight line is something cowardly drawn with a rule, without thought or feeling; it is the line which does not exist in nature. And that line is the rotten foundation of our doomed civilization. Even if there are places where it is recognized that this line is rapidly leading to perdition, its course continues to be plotted . . .

Friedensreich Hundertwasser[1]

Dose-response curves (d-r.c.s) are a traditional and widespread way to represent the analyses of data on cancer mortality in humans and experimental animals exposed to ionizing radiations.[2,3] A unique definition of a d-r.c. does not exist. In this chapter we will use a definition similar to the one given by Groer,[4] but without consideration of competing risks. The shape of d-r.c.s has been often discussed in innumerable reports and papers; however, little analytic work has been done to verify characteristic features of d-r.c.s. In this short note, we will apply theory developed by Smith[5] to data on humans[2] and beagles[3] to estimate the location of change points.[5]

DATA AND METHODS OF ANALYSIS

Data

The data on humans are from a paper by Evans.[2] He presented information on mean cumulative doses to the skeleton and on mortality from bone sarcoma and sinus carcinomas for persons exposed to ^{226}Ra and ^{228}Ra. At the end of the follow-up period of the study, 503 persons had accumulated less than one thousand rad in mean skeletal dose, and no tumors were found in this group of persons. In 102 persons with doses greater than one thousand rad, he reported a total of 43 tumors (bone sarcomas and head carcinomas).

The data on beagles are from the studies performed in the now defunct Radiobiology Division at the University of Utah.[3] In these studies, beagles were injected with varying amounts of ^{226}Ra. From the published tables[3] one can determine the sequence of tumors and other events in the order of increasing cumulative dose to the skeleton. Other events are: deaths from other causes, still alive, and lost to follow-up. If one assigns a zero (0) to other events and a (1) to bone tumors, then the whole data set can be represented as a sequence of "0"s and "1"s. Since the individual doses are not given in Evans' paper,[2] the same ordering principle cannot be applied to the human data. To circumvent this shortcoming of the human data we arbitrarily assigned the first tumor rank 507. This is the fourth position after a sequence of 503 "0"s corresponding to the persons with a cumulative dose less than one thousand rad. All tumors remained in their appropriate dose categories.

Methods of Analysis

We used a binomial model to construct the likelihood and to derive an expression for the posterior probability that the change-point occurs at position r given the sequence of outcomes $x_1, x_2, \ldots x_n$. In symbols, Smith[5] gives:

$$p_n(r) \equiv p(r|x_1, x_2, \ldots x_n) \propto p(x_1, x_2, \ldots x_n|r) \, p_0(r)$$

More specifically, the likelihood in the binomial case considered here is given by:

$$p(x_1, x_2, \ldots x_n|r, \theta_1, \theta_2) \propto \theta_1^s (1 - \theta_1)^f \theta_2^t (1 - \theta_2)^g$$

where

$$s = \sum_{i=1}^{r} x_i, \; f = r - s, \; t = \sum_{i=r+1}^{n} x_i \text{ and } g = n - r - t.$$

In the case of unknown θ_1 and θ_2 and $\theta_1 < \theta_2$, Smith[5] shows that the posterior probability is given by the following expression:

$$p_n(r) \propto Be(t + 1, g + 1)Be(s + 1, f + 1) - \Gamma(t + 1)\Gamma(g + 1)/\Gamma(t + g)$$

$$\times \sum_{i=t+1}^{t+g+1} \binom{t + g + 1}{i} Be(s + i + 1, t + g + f + 2 - i)$$

We used this expression to calculate $p_n(r)$ for the sequences of events shown in Table 13.1.

RESULTS

The results of the analyses are shown in Table 13.2 and Table 13.3. The left column in each table indicates the position r in the sequence. The right column gives the posterior probability $p_n(r)$ for the location of a change-

Table 13.1. Event Sequences for ^{226}Ra-Data in Humans and Beagles

Humans:
    506("0"s), 507("1"),508("0"),509→605[42("1"s) and 54("0"s)]

Beagles:
    33("0"s), 34("1"), 35→40["0"s], 41("1"), 42→61["0"s], 62("1"), 63("0"), 64("1"),
    65→84[15("0"s), 5("1"s)], 85→116["1"s]

Table 13.2. Posterior Change-Point Probabilities for the ^{226}Ra-Data in Humans

r	$p_n(r)$
504	0.322
505	0.565
506	1.0
507	0.005
508	0.008

Table 13.3. Posterior Change-Point Probabilities for the ^{226}Ra-Data in Beagles

r	$p_n(r)$
82	0.002
83	0.038
84	1.0
85	0.124
86	0.120

point. The probabilities were all normalized, setting the highest probability equal to one. The $p_n(r)$ agree with intuition. The highest probability for a change-point occurs always just before the majority of tumors occurs in the sequence of events. At this point θ_1 changes into θ_2, according to the specified model. These parameters can be estimated from the available data.[5] For $\theta_1 \ll \theta_2$, the change-point can be interpreted as the start of a threshold.

ACKNOWLEDGMENT

This work was supported in part by the Office of Epidemiology and Health Surveillance of the U.S. Department of Energy under Contracts Nos. 20802401 and 13512401 with Argonne National Laboratory.

REFERENCES

1. Peitgen, H.O., and P.H. Richter. *The Beauty of Fractals*, (Heidelberg: Springer-Verlag, 1986).
2. Evans, R.D., "Radium in Man," *Health Physics*, 27: 497 (1974).
3. Research in Radiobiology, Radiobiology Division Annual Report, University of Utah, C00-119-262, 1986.
4. Groer, P.G., "Dose-Response Curves and Competing Risks," *Proceedings of the National Academy of Sciences*, 75:4087 (1978).
5. Smith, A.F.M., "A Bayesian Approach to Inference About a Change-Point in a Sequence of Random Variables," *Biometrika*, 62:407 (1975).

CHAPTER 14

Radiation-Induced Carcinogenesis: Paradigm Considerations

James E. Trosko, Department of Pediatrics and Human Development, Michigan State University, East Lansing, Michigan

INTRODUCTION: DOES RADIATION REALLY "CAUSE" CANCER?

This lead rhetorical question is meant to focus on the word, "cause," since it begs us to understand the underlying mechanisms of the complex carcinogenic process and the ways ionizing radiation could contribute to these processes. To answer that question, hypotheses and theories have to be based on empirical data from human epidemiological studies, as well as experimental molecular, cellular, and whole animal studies from both radiation and chemical exposed organisms. In addition, several major concepts will have to be integrated in order to start building a "biologically-based" risk assessment model for understanding the role of low level radiation exposure in carcinogenesis.

Concepts which will have to be considered and integrated include: the multistage, "initiation/promotion/progression" model;[1] the stem cell theory versus the de-differentiation theory of carcinogenesis, or the "oncogeny as partially block ontogeny" theory of carcinogenesis;[2,3] monoclonal theory of carcinogenesis[4,5] or the "disease of differentiation" theory of carcinogenesis;[6] role of genetic/epigenetic mechanisms in carcinogenesis;[7] thresholds or no thresholds in carcinogenesis;[8] and the "nature and nurture" theory of carcinogenesis. For the purposes of this analysis, the "string" that will be used to tie these concepts together will be the concept of "intercellular communication."[9] The role of mutations, cell death, and cell proliferation, as one manifestation of epigenetic mechanisms, will be considered as potential mechanisms in the carcinogenic process.[10] Finally, the concept of an organism as a "whole which is greater than the sum of its parts"[11] because of interacting hierarchical and cybernetic systems[12,13] must be the prime paradigm.

While this will not be an easy task, the absence of this perspective of this complex biological process in a multi-cellular organism is one of the reasons that our understanding of carcinogenesis has not been as rapid as it could have been. While the reductionalistic approach is clearly necessary in the understanding of carcinogenesis and much progress can be attributed to this approach, it must be balanced and integrated with a holistic approach. Meticulous understanding of one level (i.e., molecular or cellular) cannot, by itself, be the basis for extrapolation from these "facts" to explain what might happen at the whole human level. After all, as Potter has stated, ". . . The cancer problem is not merely a cell problem, it is a problem of cell interaction, not only within tissues, but with distal cells in other tissues. But in stressing the whole organism, we must also remember that the integration of normal cells with the welfare of the whole organism is brought about entirely by molecular messages acting on molecular receptors . . ."[14] By the same token, sophisticated mathematical model fitting of empirical epidemiological data cannot provide mechanistic insight as to how agents such as radiation can contribute to human cancer, especially when extrapolations from high levels are used to estimate effects at low levels.

NO ONE THING "CAUSES" CANCER

Carcinogenesis as a Multi-step, Multi-Mechanistic Process

From epidemiological, experimental animal, human genetic, and molecular oncological studies, it has become very clear that the conversion of a normal cell to a neoplastic, invasive, and metastatic cell involves multiple "hits" and is not a one-hit process. The concept of initiation/promotion and progression[1] is an operational concept, derived from empirical experimental animal studies. No mechanistic understanding is directly implied from these studies. In addition, molecular, biochemical, or cellular studies, designed to understand the underlying mechanisms of these operational stages in the whole animal, must be carefully interpreted since none of these findings involve all the processes which occur at the whole animal level (e.g., immune systems, physiological factors). Nevertheless, important clues as to the potential mechanisms can be derived from the operational initiation/ promotion/ progression studies.

The fact that initiation appears to be an irreversible event implies mutagenesis could be the underlying mechanism. Moreover, many physical and chemical agents, characterized by in vitro tests as potential mutagens, appear to be efficient "initiators" in the whole animal tests. However, conceptually, a stable epigenetic event might also be a mechanism by which initiation at the whole animal level could occur. [By definition, *mutations* would be the quantitative or qualitative alteration of the genetic information of a cell; an *epigenetic event* would be the potentially reversible alteration of the expression of the genetic information at the transcriptional, translational, or post-translational levels.[15]] Mutations are, for all practical purposes, an irreversible event. While some epigenetic events, such as those which occur during the terminal differentiation of a cell or the developmental "commitment" of a progenitor cell, are apparently irreversible, others can be reversible (i.e., cell cycle-dependent expression of certain genes).

Mutagenic and Epigenetic Mechanisms in Carcinogenesis

Much confusion has arisen in the field because of inappropriate interpretations of in vitro data concerning potential mutagens,[16,17] of operational initiation/promotion/ progression studies as mechanistic studies, and of not having clearly defined definitions of these concepts. For example, if a chemical has been shown to induce sister chromatid exchanges or thymidylate synthetase-resistance in some in vitro system, many investigators assume these results indicate the chemical must be a mutagen. Without considering the fact that these in vitro assays are indirect measures of mutations (i.e., they are indirect indicators of phenotypic measurements) and have well-known limitations which give false positives,[16,17] any misclassification will contribute to a higher order confusion when these chemicals are now tested as potential initiators or "carcinogens."[18] Such a false positive chemical is phenobarbital. In a classical initiation/promotion system, phenobarbital is a rat liver tumor promoter. However, when a rat is fed phenobarbital "alone," a few liver tumors are always found. Does this mean phenobarbital is a weak initiator? Is it therefore a weak mutagen? When one tests phenobarbital as a mutagen in any number of in vitro tests of "genotoxicity," it turns up negative in most. Because the paradigm, "carcinogens as mutagens"[19] is so strong, the appearance of the few cancers after phenobarbital treatment implies that it has to be a mutagen. Therefore, when phenobarbital is finally tested at high concentrations in a single test and there seems to be an increase in the endpoint interpreted as indicating mutations, then one is satisfied that it is indeed a mutagen. Many such chemicals have been classified as initiators, carcinogens/mutagens based on this kind of thinking. For example, TCDD, TPA, polybrominated biphenyls, DDT, cigarette smoke condensates, acetylaminofluorene, asbestos, and various nitrosamines have been classified as genotoxicants/ carcinogens either because of their ability to induce cancers by "themselves" in animal systems not experimentally-treated with other compounds or because they have turned up "positive" in some short-term test for genotoxicity.

An alternative interpretation of these observations on chemicals, such as the ones mentioned, is that they are not genotoxic but rather they promoted preexisting initiated cells. In this case if one assumes that initiation is due to a mutation, then these preexisting initiated cells might have been mutated either by spontaneous mutagenesis which can occur with a small, but finite frequency, every time a cell divides. Also, since cells are constantly exposed to environmental mutagens, particularly background radiation, one should predict that the older the organism is, the more initiated cells should be found in the organism. Therefore, in animal experiments designed to test a single compound for its "carcinogenicity," the animal is not immune from other "carcinogens" which it must also confront in its food, living environment, or its own metabolism. In addition, if one does not assume that all initiation has to be associated with a mutagenic event, then these nongenotoxic "carcinogens" might actually create an abnormal micro-environment which allows a stem cell to escape the negative growth control of the surrounding normal differentiated cells and to proliferate without differentiating into the normal cells of that tissue. Teratocarcinomas might be an example of this kind of process. By putting these teratocarcinomas into a more normal micro-environment, they could differentiate into their normal cell lineage.

In effect these "nongenotoxic carcinogens" are "mitogens" rather than "mutagens"[20,21] by their ability to bring about the clonal expansion of stem or progenitor cells which cannot complete terminal differentiation. Now, there are several ways by which promotion or clonal expansion of these cells can come about. The first is noncytotoxic stimulation of cell proliferation. Growth factors, by definition, would be examples of this sort. Chemicals, which by triggering transmembrane mitogenic signals via nonreceptor-type mechanisms such as DDT or phenobarbital, might be examples of this class. More recently, the idea that a chemical, by its ability to inhibit apoptosis,[22-26] could increase the population of cells which might otherwise have died, would be another possible mechanism to promote initiated cells.

A second means to promote initiated cells is via cell removal or cell killing. Cytotoxicity of chemicals, viruses, or physical agents could bring about the proliferation of surviving initiated cells.[27] This would be particularly true if the cytotoxic effect on the tissue was selective; that is, if the initiated cell was more resistant to the cytotoxic agent. This, of course, has been observed in cases where initiated enzyme altered cells in a focus found in rat liver could not metabolize chemicals into electrophilic forms.[28] Cell removal by abrasion, surgery, burns, etc., could stimulate compensatory hyperplasia, causing any surviving initiated cell to repopulate the tissue.[29,30]

By expanding the number of initiated cells during this promotion process, one increases the number of cells with one defect of several that are needed for the cell to become neoplasic, invasive, and metastatic. In other words, if there are several rare events needed of complete neoplastic transformation, the probability they would all happen in one cell would be the result of multiplying the rare independent probabilities of each event needed.[31,32] However, if one expands the number of cells with one of the "hits" needed, the probability that one of those cells will accrue a second, third, or more hit will be increased by the number of those cells. In effect, promotion is that process which increases the "target size" of the critical cells.[33,34]

If the promotion or clonal expansion process is interrupted or stopped before the acquisition of the hit needed to stabilize the process, the promoted population of initiated cells appears to regress or disappear.[35] Obviously, not all the initiated cells disappear, since the initiated tissue, if re-promoted, will give rise to papillomas in the skin or foci in the liver. These observations have led to the idea that promotion is an interruptable or reversible process. This, in itself, would suggest that promoters are not mutagens. The interruption of the promotion process and the disappearance of these clones of initiated cells might be related to apoptosis. As long as promoters are present, apoptosis is prevented and these initiated cells can increase.[22-26]

The process, by which an event in an initiated cell allows that cell to become promoter-independent and that is independent of an external mitogenic stimuli, could be defined as the conversion or progression step. Agents which could bring about the necessary phenotypic characteristics needed for autonomous growth could be either genotoxic or a stable epigenetic in nature. The concept of initiation/ promotion/initiation was designed to test this idea, and the evidence to date seems to support it.[36-43] It would seem clear that the underlying mechanisms for each of these steps, e.g., initiation, promotion, progression, are very distinct and different. That which could cause an irreversible event would be quite different from that which can be mitogenic. In fact, if one assumes for the moment that mutagenesis can initiate a cell, while promoters can stimulate a cell to proliferate, then there are probably multiple ways one can mutate cells, as well as multiple ways in which cells can be stimulated to divide.

STEM CELLS, INITIATION, AND PARTIALLY BLOCKED DIFFERENTIATION

Several important concepts in carcinogenesis must be considered in the context of the initiation/promotion/progression theory of carcinogenesis; basically, the observation that cancer cells appear to be the clonal derivatives of a single cell, which have proliferative potential, yet have, depending on the tumor, various degrees of differentiation of the tissue lineage of which they are a part. While the issue of whether the cancer is the result of a stem cell which has been partially blocked in its ability to completely differentiate[3], or the result of a differentiated cell having been forced to dedifferentiate is still unresolved, an assumption will be made for the purpose of this analysis; namely, most, if not all, cancers originate from a stem or committed progenitor cell.[44,45] While there are several lines of evidence consistent with this hypothesis, the observation that only a few contact insensitive cells in a population of normal cells derived from an embryo can be transformed[46] will serve as a major line of evidence. That this stem cell theory has relevance to risk assessment is based on the following: (a) not all the cells of the body are target cells for the carcinogenic process, but only the few stem cells; (b) the number of stem cells will vary in different tissues and in some tissues decrease with age and, in the case of breast tissue, be influenced by the pregnancy status; and (c) initiated stem cells, on the other hand, might be expected to increase with age and exposure to tumor promoters.

If stem cells are the target cells, then initiation must start with these cells. Consequently, one can now speculate that the biological consequence of the initiation process is to block the differentiation process without blocking the ability of this stem cell from proliferating (Figure 14.1). As long as the stem or committed progenitor cell is inhibited from dividing, this initiated cell can remain quiescent for long periods.[35] When it is stimulated to proliferate by either noncytotoxic or cytotoxic means, it will accumulate in the tissue as abnormal, partially differentiated cells of that tissue lineage . . . the state of differentiation being dictated by which genes have been altered by the initiation process. Normal stem or committed progenitor cells receiving the same mitogenic stimuli would also proliferate and terminally differentiate. As a result, the proliferation or promotion process would bring about a differential accumulation of these two types of cells. In effect, the initiated or partially differentiated cells would selectively accumulate, since

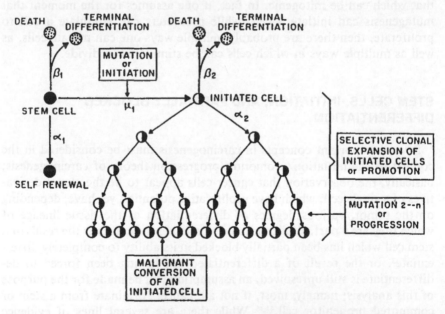

Figure 14.1. The initiation/promotion/progression model of carcinogenesis. β_1 = rate of terminal differentiation and death of stem cell; β_2 = rate of death, but not of terminal differentiation of the initiated cell (-||→); α_1 = rate of cell division of stem cells; α_2 = rate of cell division of initiated cells; μ_1 = rate of the molecular event leading to initiation (i.e., possibly mutation); μ_2 = rate at which second event occurs within an initiated cell. (From Trosko et al., with permission from Alan R. Liss, Inc., New York).

in most tissues the terminally differentiated cells would eventually die or be removed. Alternatively, if the promotion process blocks apoptosis and apoptosis is the means by which certain cells, such as initiated ones, are removed, then in time they would accumulate.[22-26]

In certain tissues, such as the lining of the G.I. tract, or skin, the number of stem cells might be expected to remain relatively constant throughout life and to have relatively high rates of cell proliferation. Therefore, the number of target cells would be expected to be relatively large and the probability for mutagenesis/initiation to be enhanced by the high probability of a DNA lesion to be fixed into a mutation by cell division. The cell division itself could be a promoting factor. This might be the explanation for the relatively high cancer rates associated with the skin and parts of the G.I. tract. The fact that high rates of cell division occur in the small intestine but few cancers appear there challenges this hypothesis,[47,48] unless (a) the progenitor cells of the small intestine are far down the differentiation pathway, preventing multiple hits from occurring before the cell is sloughed off or (b) strong naturally occurring antitumor promoters keep these initiated cells from proliferation.

This, then, raises the question related to species, tissue, and cell type specificities of initiators and promoters. With regard to chemicals, the dynamics of tissue distribution and metabolism would be expected to play a major role for both initiators and promoters. For physical initiators, such as radiations, these factors would not be important. Only those factors influencing the direct or indirect effect of the photons, ions, or radicals might be expected to be important. Species and tissue specificities have been well-documented for both chemical initiators and promoters.[49,50] On the other hand, the cell type specificities for initiator or promoter action can only be inferred. For example, if the biological consequence of tumor promotion (chemical or physical) is mitogenesis, then one can see that some promoters, such as phenobarbital or polybrominated biphenyls, seem to bring about the selective proliferation of initiated cells but not the normal cells in the liver. There is not a general sustained hyperplasia of the total liver tissue. On the other hand, 12-O-tetradecanoylphorbol-13-acetate (TPA) treatment of the Sencar mouse skin appears to cause a general sustained hyperplasia in the tissue in which it causes promotion.[51]

NATURE AND NURTURE OF CARCINOGENESIS

How many times has it been asked whether it was the genes or the environment which "caused" any cancer? There was a time in the not too distant past where few believed that genetics played a significant role other than in a few rare syndromes, and in spite of early observations of chromosomal anomalies associated with tumor cells.[52] The skepticism should now be mute with a wide variety of evidence, particularly from molecular oncology (see next section). However, to illustrate how environmental and genetic factors interact and how this concept can be integrated into the previous concepts, several genetically-predisposed cancer prone syndromes will be examined in light of the initiation/promotion model of carcinogenesis.[53]

In principle, the underlying mechanisms to the initiation and promotion phases are very different, one being a very stable or irreversible event (possibly involving the stable epigenetic or mutagenic alteration of a single stem cell), and the other being a reversible or interruptable process affecting the clonal expansion of that single initiated cell by a mitogenic mechanism. It should be obvious that both endogenous and exogenous mutagens could interact with the DNA of cells. In addition, it should also be obvious that there are endogenous and exogenous chemicals which could protect the DNA from these mutagens. They would be conceptualized as "anti-initiators." Genetic and environmental chemicals could either protect or enhance the chance of an initiation event to occur. Sun screens or drug metabolizing/inducing chemicals could, for example, reduce the damage to DNA caused by ultraviolet light in the skin or hepatotoxicants damage in the hepatocytes, respectively. Genetically-determined pigmentation of the skin (albinos, caucasians, orientals, blacks) would influence the amount of ultraviolet light damage to DNA.[53] These genes then would be considered initiator-influencing genetic factors.

Once the DNA has been damaged, conversion of these lesions to an altered genome which characterizes the initiation event can be influenced by the repair or lack thereof to these lesions. Again, both genetic and environmental factors could influence this phase of the initiation process. Any exogenous or endogenous chemical which could enhance or inhibit the repair of these lesions could increase or decrease the initiation event. Xeroderma pigmentosum, a genetic syndrome which predisposes the individual to sunlight-induced skin cancer, is an illustration of an initiation and promotion-prone syndrome, in that by the inability to repair sunlight-induced pyrimidine dimers, both mutation (i.e., initiation) and cell death [promotion via compensatory hyperplasia[54]] can occur.[55,56] Bloom syndrome is another cancer-prone syndrome in which the spontaneous mutation rate is elevated by some as yet unknown mechanism.[57] Chemicals which might specifically inhibit or stimulate DNA repair could, in principle, enhance or reduce the initiation phase of carcinogenesis.

Using the same kind of thinking, there are endogenous and exogenous factors influencing the promotion phase of carcinogenesis. Examples of chemicals which have been shown to be promoters are the exogenous natural plant toxic chemicals (e.g, phorbol esters), pesticides, herbicides (DDT, 2,4,5-T), pollutants (polybrominated and polychlorinated-biphenyls), dietary ingredients (unsaturated fatty acids), food additives (saccharin and butylated hydroxytoluene), cigarette tar condensates, and drugs (valium, phenobarbital), growth factors, and hormones.[58] Genetic diseases which would increase or decrease chemicals, such as growth factors or hormones, could stimulate or inhibit the promotion process.[59-67]

The recent identification of the genes affecting several cancer-prone genetic syndromes, such as retinoblastoma and Li-Fraumeni,[68,69] could give further clues as to how genetic factors can influence the initiation or promotion phases.

ONCOGENES, TUMOR SUPPRESSOR GENES, AND THE MULTISTAGE NATURE OF CARCINOGENESIS

Assuming now that the evidence for the initiation/promotion/progression theory and the definition of an initiated cell as a stem cell which is unable to terminally differentiate are accepted, then the next question to be examined is "What are the genes which are responsible for the initiating and promotion events/processes?" In the last decade, due largely to advances in molecular/ cell biology, the concepts of "oncogenes" and "anti-oncogenes" or tumor suppressor genes have emerged.[70-72] By definition, proto-oncogenes are those sequences of DNA which code for growth factors, transmembrane signaling elements, and nuclear transcription factors influencing cell proliferation and differentiation.[73] Conversion of a proto-oncogene of a normal cell to an activated oncogene associated with a tumor cell can be brought about by a mutagenic event, gene amplification, or abnormal gene expression process. Early interpretation of oncogene experiments suggested at least two independent events had to occur in the standard NIH-3T3 transformation assay.[74,75] Later studies made this neat story more complex, in that in some transformation assays only one hit seemed necessary,[76] or six to eight events needed to occur before a given cell type was transformed to a fully tumorigenic phenotype, under the conditions of the particular study.[77,78]

The definition of a tumor-suppressor gene is an operational one which implies that it could prevent the ultimate formation of a tumor at any of the multiple steps. Operationally, several classes have been identified.[79] In principle, any gene that would be considered an anti-initiator, as well as an anti-promoter could be conceived as a tumor suppressor gene. DNA repair genes, by reducing the probability of mutational events, could be viewed as tumor suppressor genes. Genes negatively controlling cell proliferation or positively controlling cell differentiation also could be viewed as tumor suppressor genes.

If initiation is the result of a stable alteration of a gene in a stem cell which allows the cell only to partially differentiate but does not prevent its ability to proliferate, then the question is, "Does initiation involve the proto-oncogene or the tumor suppressor gene?" If one accepts the aforementioned definitions and concepts, it seems clear that this might be a meaningless question, since both proto-oncogenes and tumor suppression definitions could influence the differentiation process.

One of the major spin-off concepts involved in the multistep carcinogenic process is the hypothesis that the process involves the conversion of a normal, mortal stem cell into an immortal, tumorigenic cell. In other words, the first step of this process must involve the immortalization process, then the malignant transformation can occur next.[74,75] An alternative view is that the stem cell, which is speculated to be the target cell for transformation is, by definition, immortal. It must, for self renewal potential, remain undifferentiated. Once one of the stem cell daughters has been induced to become committed to terminally differentiate rather than maintain its self-renewal potential, it has been "mortalized." Therefore, if initiation is the process by which a stem cell, which is immortal, is altered so that it cannot terminally differentiate or become mortal, then our concept has been wrong. The first process that must occur during the initiation phase of the multistage nature of carcinogenesis is the prevention of mortalization of the immortal stem cell, rather than the immortalization of a mortal cell.

GAP JUNCTIONAL INTERCELLULAR COMMUNICATION: AN INTEGRATING LINK IN THE UNDERSTANDING OF CARCINOGENESIS

Intercellular Communication, Homeostasis, Growth Control and Differentiation

If a cancer cell is characterized as a cell which has lost growth control and cannot terminally differentiate, the understanding of the carcinogenic process must focus on those genes and cellular processes needed to regulate these two important cellular functions. From an evolutionary perspective, since cancer is only a property of a multicellular organism where the control of multiple cells within the organism and the need to regulate differentiation during the development of that organism must be carefully orchestrated, the search for those new genes responsible for this function must include the thought that it might be related to the cancer process. To affect this coordinated and orchestrated process, a series of intercellular communication mechanisms has evolved, apparently in a highly integrative fashion. The mechanism by which two noncontiguous cells can communicate is extra-cellullarly by means of growth regulators, hormones, and neurotransmitters.[80-83] The target cells for these extracellular signals respond by having a set of intracellular signals affect posttranslational, translational, and transcriptional regulatory elements within the cell. This, in turn, can now affect the relationship continuous cells have with each other by means of gap junctional intercellular communication channels [Figure 14.2].[21] In other

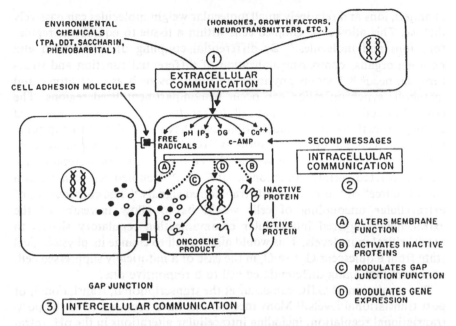

Figure 14.2. The heuristic schemata characterized the postulated link between extracellular communication and intercellular communication via various intracellular transmembrane signalling mechanisms. It provides an integrating view of how the neuroendocrine-immune system ("mind or brain/body connection") and other multi-system coordinations could occur. While not shown here, activation or altered expression of various oncogenes (and "anti-oncogenes") could also contribute to the regulation of gap junction function. [Reprinted from J.E. Trosko and C.C. Chang, *Toxicology Letters*, 49:283–295 (1989), Elsevier Science Publishers; used with permission].

words, the orchestrated control of cell proliferation, cell differentiation, and adaptive responses of differentiated cells within a multicellular organism is mediated by the highly integrated systems of these three major forms of cell communication (cell matrix, cell-cell adhesion mechanisms are viewed for this analysis as a subset of intercellular communication). These three forms of communication constitute the physiological basis of homeostasis. Extracellular communication can modulate the functions controlled by intercellular communication through perturbations of intracellular communication signals in the target cells. By the same token, modulators of intercellular communication can affect the levels of intracellular communication signals, which then could affect extracellular signal process.

It is interesting to note that during the evolutionary transition from single-celled to multi-celled organisms, the gene for intercellular communication appeared; namely, the gap junction gene.[84] The gap junction is a membrane-associated protein channel between contiguous cells, composed of six proteins (connexins) in a hemi-channel, (connexon). Through these channels, ions and relatively small molecular weight molecules can passively diffuse. This allows the coupled cells within a tissue to equilibrate regulatory ions and molecules.[85] By differential coupling of cells within and between organs, compartmentalization of differential function and structure can occur.[86] Various physiological functions such as electrotonic and metabolic synchronization can occur in compartmentalized regions. The control of cell proliferation and cell differentiation, as well as specialized adaptive functions of communication cells, have been linked to gap junctional intercellular communication (GJIC).[87,88] The synchronization of heart[89] and uterine cells,[90] as well as the control of insulin secretion have been associated with the functioning of GJIC.[87] Coupled cells can act as a "point source" or "sink" for GJIC transferred signals.[91] The consequence of extracellular uncoupling of cells would be to allow increases of the extracellular-triggered intracellular communication regulatory signals to reach critical mass levels. This would allow a cell to change its physiological state from a quiescent G_0 to a G_1 in the case of a mitotically suppressed cell, or a nonfunctioning differentiated cell to a responsive one.

The regulation of GJIC can occur at the transcriptional, translational, or post translational levels.[92] Many mechanisms have been linked to the post-translational regulation, including intracellular alterations in the pH, intracellular free calcium, C-AMP, as well as phosphorylation of the gap junction protein.[93] Other mechanisms also have been postulated.[94] Most cells of a multicellular organism have gap junctions at some time in their development. Only a few free-standing cell types, such as certain neurons, free-floating lympho-reticular cells, and early embryonic cells do not express the gap junctions. In solid tissues most cells are coupled and contact-inhibited. While rigorous evidence associating contact inhibition and GJIC is still lacking, there is strong circumstantial evidence linking GJIC with contact inhibition.[88] During cell removal, cell death, or wounding, cells seem to lose their GJIC during the regeneration process.[95,96] Most, if not all, cancer cells which do not contact inhibit have decreased or altered GJIC.[97-99] Many mitogens, such as epidermal growth factor can down regulate GJIC.[81,82,100] Free-floating progenitor lympho-reticular cells are regulated by various extracellular communication signals,[101] Yet there is evidence that in their maturation process, GJIC may be required for the completion of their differentiation.[102,103] Stem cells, such as the totipotent fertilized egg[86] and pluri-potent stem cells of the kidney and breast, appear not to have any GJIC.[104,105] These cells also are probably mitotically regulated by negative extracellular growth signals from their differentiated daughters.[106-109]

GJIC, Dysfunctional Growth, and Differentiation Regulation and Carcinogenesis

There exists a wide variety of experimental evidence linking abnormal GJIC and carcinogenesis[110] To begin, most normal cells of solid tissues have gap junctions and are contact inhibited. Normal stem cells or mature red blood cells and many terminally differentiated neurons do not. Since these terminally differentiated cells will never divide, there seems to be little functional need for them, even though they may be present in the early maturation process of these cells.[102,103]

Most, if not all, cancer cells have dysfunctional GJIC.[97-99] Even though "exceptions to this generalization have been made,"[111] because of the nature of these studies (i.e., detection of the gap junction protein in some cancer cell does not indicate their functionality), these exceptions do not rigorously negate the hypothesis that cancer cells have defective GJIC. Many studies have shown either selective communication[112,113] or a decrease or down regulation of GJIC in various tumor tissues.[85,114-117] In other words, the relationship of GJIC and cancer could be quantitative, as well as qualitative.

Most, if not all, endogenous and exogenous tumor promoters can reversibly down regulate GJIC. One of the operational characteristics of tumor promotion in vivo is the reversibility of the process. Both in vivo[118] and in vitro, a wide variety of tumor promoting conditions (i.e., wounding[95,96]), endogenous growth factors,[81,82] or hormones[80] and exogenous chemicals[58] have been shown to reverse down regulate GJIC at the transcriptional, translational, and posttranslational levels. In those few cases where an in vivo promoter has not been shown to inhibit GJIC in vitro (i.e., TCDD),[119] the problem of the vitro system not mimicking the in vivo environment could be the explanation. In addition, there is also the possibility that the chemical would only modulate GJIC in the initiated cell, and not the normal cells being tested in vitro.

The fact that many oncogenes have now been shown to be associated with the down regulation of GJIC provides strong evidence that GJIC has something to do with carcinogenesis. A series of activated, transforming oncogenes (e.g., Src, Ras, Neu, Raf, Mos) have been shown to down regulate GJIC in a variety of cells.[120-133] Transformation studies have shown that cooperation of several oncogenes (i.e., myc and ras) were needed for full transformation. Reduced GJIC and tumorigenic growth were correlated in raf transformed cells; however, a myc- and raf-transformed cell showed dramatically reduced GJIC and enhanced tumorigenicity.[133] In addition, when ras-transformed cells, which have reduced GJIC, are treated with TPA, GJIC is reduced to zero and the tumorigenicity is increased synergistically.[126]

Given that tumor suppressor genes prevent, negate, or ameliorate the effects of oncogenes, one would expect that if oncogene effects are mediated, in large part, by the down-regulation of GJIC, then one would predict the tumor suppressor genes to up-regulate the GJIC. This appears to be the case. Lee et al.[134] and Kalimi et al.[135] have shown that several tumor suppressor genes are associated with GJIC. The molecular/biochemical mechanism by which these tumor-suppressor genes might act could be at several levels. For example, if a given oncogene product brings about the phosphorylation of a gap junction protein, then a tumor suppressor gene might bring about rapid de-phosphorylation. The fact that one chemical tumor promoter inhibits the phosphatase activity of a cell would indicate that regulation of phosphorylated proteins, such as the gap junction proteins, would be critical in cell regulatory processes.[136]

Anti-carcinogens and anti-tumor promoters, such as retinoids, carotenoids, c-AMP, green tea components, have been associated with the up-regulation of GJIC.[137-148] While the mechanisms by which these diverse chemicals negates the tumorigenic or tumor promoting processes are unknown at present, one would surmise that they act via different mechanisms.

Anti-oncogenic drugs, such as Lovastatin, specifically reverse the down regulation of ras-down-regulation of GJIC.[149] In this case, the H-ras oncoprotein is only functional when it is bound to the undersurface of the plasma membrane.[150] It has been shown that lovastatin prevents that process.[151] Lovastatin can prevent the growth of ras tumorigenic cells in nude mice.[152] This, then, is another example of the reversal of tumorigenicity with the up-regulation of GJIC.

Recently, several noncommunicating tumorigenic cells were transfected with gap junction genes.[153-156] In these cases, GJIC was restored and a more normal regulation of cell growth control was noted, as well as a slowdown of tumorigenicity. The failure to completely restore nontumorigenicity could be due to many reasons, not the least of which endogenous factors down-regulated the gap junction function in vivo.

If GJIC is necessary for normal growth control, then mutants for this function should be tumorigenic. Oh et al.[157] have recently shown that rat liver epithelial cells, mutated for GJIC, were not only total defective for GJIC, but also were tumorigenic when placed into the rat liver. What was interesting was the fact that, even though there was absolutely no GJIC in these cells, the time it took to become a tumor and the tumor size was longer and smaller, respectively, than oncogene-transformed rat liver cells which have only reduced GJIC. This might imply that GJIC is a necessary, but insufficient, factor in a cell's ability to become tumorigenic. In other words, an activated oncogene in cooperation with another oncogene or the loss of inhibiting tumor suppressor genes would be able to block GJIC, while also stimulating the mitogenic process. On the other hand, a specific mutation which only affects the function of GJIC but not the other coordinated mitogenic steps would still be unable to divide without other exogenous mitogenic stimuli.

One of the critical tests of the role of GJIC in carcinogenesis would be the use of anti-sense gap junction genes to negate the function of GJIC in normal, growth controlled and GJIC-proficient cells. Experiments are being performed to test this hypothesis.

While all of these aforementioned studies correlate with the down- or up-regulation of GJIC with tumorigenicity or normal growth control, respectively, all that can be said is that they are consistent with the hypothesis that GJIC plays a role in carcinogenesis.

INSIGHTS FROM RADIATION AND CHEMICAL CARCINOGENESIS: IMPLICATIONS FOR RISK ASSESSMENT

While very early studies in radiation genetics gave us the fact that radiation could induce mutations,[158] and early studies in cancer gave clues as to the role of mutations in cancer,[52] the impact of the studies of radiation and chemical carcinogenesis is only now beginning to take place. It seems that, while there has been some cross-fertilization between these two fields, some important observations and concepts of each field have not influenced the other significantly. The concept of DNA repair and its link to carcinogenesis via increased mutagenesis came from the DNA repair deficient, skin cancer prone syndrome, xeroderma pigmentosum.[55,56] When the chemical carcinogen, acetylaminofluorene, was shown to induce a DNA lesion which was not repaired in xeroderma pigmentosum cells,[159] a merging of radiation carcinogenesis and chemical carcinogenesis started to come about. The multistage nature of carcinogenesis came from the field of chemical carcinogenesis.[160] However, this concept has not been a predominant paradigm in the field of radiation carcinogenesis. Experimental findings from both fields could give important insights to the mechanisms of carcinogenesis relevant to risk assessment.

One example is the fact that, while all cells of xeroderma pigmentosum individuals lack DNA repair, most of the cancers found in these individuals are in the skin.[161,162] Since the skin is the target tissue for the most prevalent environmental mutagen, ultraviolet light, and since the amount of DNA lesions would be the same for equally exposed normal and xeroderma pigmentosum individuals, it is not surprising that risk to ultraviolet-light-induced skin cancers is higher in xeroderma pigmentosum than normal individuals. However, these individuals are also exposed on a daily basis to thousands of chemical carcinogens, particularly via the diet. Yet, the frequency of G.I. and liver tumors is surprisingly low if we assume these chemicals are primarily mutagens, and if we ignore the multistage nature of carcinogenesis. In fact, the high frequency of skin tumors in these individuals has to conform to the multistage nature of cancer. Cell killing by ultraviolet light is probably an indirect promoter.[54] The amounts of chemicals to which any tissue of xeroderma pigmentosum is exposed are obviously below cytotoxic levels. Therefore, any initiated cell of internal tissues must be promoted and converted in a manner similar to normal individuals. This should emphasize the point that DNA damage and mutagenesis (i.e., initiation), while a necessary event, is insufficient for carcinogenesis.

While the multistage nature of carcinogenesis was derived from the field of chemical carcinogenesis, it has been generally assumed that the chemicals used to initiate the process are, in fact, DNA-damaging agents and muta-gens.[19,163] This is based on the fact that the initiation appears irreversible, the chemicals used, in some cases, were positive in some short-term tests for genotoxicity, and mutations were found in oncogenes of tumors of these chemically exposed animals. However, these assumptions have to be veri-fied since there are serious questions which need to be explained and answered. First, just because a chemical seems to initiate a tumor in an animal, it might do so via some stable epigenetic mechanism. It is now also very clear that all in vitro "genotoxicity" assays have serious limitations leading to far too many false positives.[16,17] There is also strong evidence that chemicals used to induce the cancers do not seem to have been responsible for the induction of the mutations found in the oncogenes of the transfor-mants or tumors,[164-173] but rather the chemicals seem to have promoted preexisting, spontaneously existing initiated cells.[174] All of these dilemmas call into question not the initiation/promotion/progression concept, but rather the assumption that chemicals, which are associated with cancer production in certain strains of rodents, are acting via DNA damaging and mutagenic mechanisms. Even the fact that DNA lesions are found after DNA, cells, or animals are exposed to chemicals must be carefully inter-preted. It is not surprising that reactive electrophiles will react to DNA in cell-free in vitro experiments. It is also not surprising that one can find DNA lesions or mutant or phenotypically altered cells or organisms after exposure to electrophilic chemicals, or those which can be metabolized in certain cells (i.e., hepatocytes).

This latter case brings all of the previous concepts to bear on the under-standing of carcinogenesis. Most of the bioassay evidence of chemical car-cinogenesis comes from rodent assays, in which certain strains and the liver seem to be the primary source of those data.[175] When one isolates DNA from treated livers or looks for DNA repair in hepatocytes or measures mutations in nonmetabolizing cells with liver microsomes, it is not surpris-ing that one sees evidence of DNA lesions, DNA repair, or finds mutations in the microsome-treated cells. However, are these the cells which give rise to cancer? Are the highly differentiated hepatocytes the target cells for liver carcinogenesis, or are the nonmetabolizing stem cells the target cells?[176] Are these hepatocytes simply "kamikaze" or "drone" cells, evolutionarily designed to protect the stem cells of the liver from toxic chemicals? If the stem and differentiated hepatocytes were of equal sensitivity to environ-mental chemicals, the liver, hence, organism, would not survive and regen-erate after a cytotoxic exposure. The ability of metabolic toxicants pro-duced in the hepatocytes to be transported to the target stem cells would be highly unlikely at low doses, especially for highly reactive electrophiles. At high doses, the damage to the hepatocytes would lead to cytotoxicity of the hepatocyte. This could then be the stimulus for inducing a compensatory or regenerative hyperplasia leading to promotion of previous spontaneous or induced initiated cells. Whether radiations, which do not have to be metabolized to damage DNA, bring about a differential response in stem or in differentiated daughter cells is unknown, although the physiological state of the cell surely could influence the primary damage or its repair.

The concept of the multistage nature of carcinogenesis has been only infrequently used in the design of experimental radiation carcinogenesis and in the interpretation of human radiation cancer epidemiology.[177] The first question which still needs to be resolved is, "Is ionizing radiation an initiator, promoter, or progressor of carcinogenesis?" Since it is a fact the exposure to both acute and chronic radiations is associated with the appearance of cancers in both experimental animals and in humans exposed to therapeutic or A-bomb radiation,[178-183] as well as transformation of cells in vitro,[184-186] the mechanism by which radiations can affect any of these steps in carcinogenesis needs to be examined.[187]

While it is also a fact that ionizing radiation can induce mutations in cells, the nature of those mutations appears to be primarily gene deletions, chromosome deletions, and chromosomal rearrangements.[188-193] In fact, the proportion of deletions is probably larger than measured by conventional methods, since most are lost when the cells die.[194] Ionizing radiation appears to be relatively ineffective as an inducer of point mutations, especially as evidenced by the survey of point mutational changes as detected by altered protein mobility coded by genes of the offspring of A-bomb survivors.[195] Only sophisticated analysis of these individuals by modern molecular DNA technology might resolve this issue.[196] One area which has received virtually no attention is the possibility that ionizing radiation could produce "epigenetic" alterations by either chromosomal rearrangements or by demethylation of genes. In both of these cases, if either one or both of these events could occur, the genes, while not mutated, might be expressed in an abnormal manner. Clearly, ionizing radiation can be a cytotoxicant.

How might the understanding of the molecular nature of ionizing radiation induced lesions help in understanding the role of radiation carcinogenesis and in risk assessment? In the context of the initiation/promotion/progression theory, the few examples, while suggesting ionizing radiation could be an initiator, promoter, and progressor,[41,197,198] still, the studies were not unequivocal. As a potential initiator, the persistent problem of radiation causing the indirect promotion of preexisting initiated cells by its cell killing effects cannot be ruled out. In the context of the oncogene/tumor suppressor gene theory, proto-oncogenes need to be activated either by mutations, abnormal gene expression, or gene amplification. Ionizing radiation could conceivably activate a proto-oncogene by a deletion mutation which might truncate a growth factor receptor, for example, or by an infrequent point mutation. In addition, since ionizing radiation can induce chromosomal rearrangements, the stable translocation of an oncogene to a new chromosomal site could stably de-repress a dormant proto-oncogene. Conceivably, if ionizing radiation can cause de-methylation of genes leading to either a stable "imprinting"[199,200] or de-repression of the proto-oncogene in either germ or somatic cells, an activated oncogene might be the result.[201-203] In addition, several observations demonstrating transmission of chromosomal instability and delayed reproductive death in irradiated cells could be consistent with an epigenetic event causing a conditional mutator or lethal event.[204,205] It should be obvious that the answer to the question as to ionizing radiations' ability to initiate and to cause de-methylation of DNA is still unknown.

Can ionizing radiation be a promoter? If one assumes that mitogenesis is responsible for the clonal expansion of initiated cells, cell killing induced by radiation, depending on the amount and the kind of cells killed (stem, differentiated), could act as a stimulus for regenerative hyperplasia. It has been reported that low levels of low energy transfer (LET) radiation can induce protein kinase C message and stimulate protein kinases in cells.[206-208] Since protein kinase C is intimately involved in tumor promotion[209] and in the inhibition of GJIC,[210] there is a possibility that ionizing radiation can, at low doses, act to clonally multiply initiated cells without cell killing.

The possibility that ionizing radiation might be a "progressor" of the multistage carcinogenic process would seem to be an easy answer to obtain experimentally. However, in theory, based on the oncogene/tumor suppressor gene hypothesis and the observation that most of the mutations induced by ionizing radiation seem to be deletions, the loss or inactivation of tumor suppressor genes might be predicted to be associated with the late stage of carcinogenesis,[187] and consistent with the loss of several tumor suppressor gene alleles found in many tumors.[211]

Clearly, the interpretation of cell killing, mutagenesis, and cell transformation data, associated with the quantity and quality of radiations (dose, dose rate, high -, low LET) on in vitro systems to explain experimental animal and epidemiological human data, has to be done with great care. Many artifacts are associated with the interpretation of in vitro cytotoxicity, mutagenicity, and transformation results. To begin with, when one irradiates cells in an organism, most cells exist, not as individual discrete units, but instead are coupled into a syncytium. It is the syncytium which is the "target" of the radiation energy. Clearly, at the molecular level, DNA is considered the critical target. Yet the final consequence of the radiation effect on that DNA is affected by the coupled cells, which could provide protection either by quenching some of the ionizing-related DNA damage, by suppressing fixation of lethal or premutagenic DNA lesions of contact-inhibited cells because these cells are not in cell division, and by allowing more time for DNA repair. Evidence suggesting that cell coupling might play a role in radiation protection comes from studies where the same kind of cells are irradiated in dense or sparse situations, or in contact inhibited versus noncontact inhibited conditions.[212-219]

Radiation studies, using in vitro transformation systems, usually start with cells which are not normal, primary, or diploid cells. In many cases, a single acute exposure will lead to morphological transformants. In others, a "two-stage-type" transformation phenomenon is seen. However, it cannot be rigorously assumed that the system mimics the in vivo situation. The fact that many of the chemical and physical carcinogens used to induce these "mutants" and transformants easily lose their transformed phenotype,[220,221] are induced by chemicals which act via nonmutagenic mechanisms,[222-224] and are induced at relatively high levels which do not induce many mutations at the same time in the same cells,[225,226] leads one to suspect that these morphological transformants are the result of some epigenetic mechanism.[227] The fact that azacytidine can modulate transformation cells and not induce mutations,[43,228,229] while at the same time alter the methylation patterns of the DNA of the treated cells and cause differentiation,[230-232] provides strong support to this idea. Studies have shown altered patterns of methylation in tumor versus the normal tissue.[233,234] The high frequency of "initiation" events as measured in some in vitro transformation assay[235] might be the result of radiation induced "epigenetic," not mutagen events.[227] The differences seen in various biological effects of low and high LET studies might reflect a real difference in the abilities of these two forms of radiation to bring about these kinds of events.[236] The fact that high LET radiation from α-particles of plutonium 238 induced a significant increase of sister chromatid exchanges (SCEs) compared to x-rays (0.31 mGy versus 2.0 Gy), only 1% of the cell nuclei in the cells were transversed by any α-particle.[237] Given that SCEs can result from artifacts,[238] these effects may reflect α-particle induced indirect effects transmitted by mitogenic factors released from dying cells. Very few studies have been done to test if ionizing radiations can induce altered methylation of genes.

When in vitro mutation assays are done using selection for altered phenotypes (i.e., drug-resistance), the type of "genetic marker" used for studies of the effects of different qualities and quantities of radiations appears to suggest gene-dependent responses.[236] While it might be an overstatement, it seems that DNA in DNA and gene dependent effects might be predicted to be minimal. Alternatively, since these "gene markers" are really phenotypic markers, not genotypic markers, some do, in fact, measure epigenetic alterations of the gene, not mutagenic changes.[16,17] For example, the thymidylate kinase or resistance to bromodeoxyuridine marker and the hypoxanthine guanine ribotransferase or 6-thioguanine resistance marker could detect point mutations which knock out the enzyme activity associated with those genes, intergenic and gross chromosomal mutations, and epigenetic repression of those genes. On the other hand, the Na/ATPase or ouabain resistance marker, if epigenetically repressed or mutationally deleted, would lead to the death of a cell carrying such an alteration.[239] If ionizing radiation, either a particular quality or quantity, primarily induces deletion mutations, there is a great probability that other vital genes might also be lost. Therefore, the measured frequencies in vitro might be minimal estimates.[240] This has recently been verified using a human female cell line, which utilizes the trick of inactivating the active X chromosome and activating the inactive X chromosome.[194] This demonstrated that the majority of the cells containing such deletion mutations are lost in normal cells. While this is clearly not an exhaustive analysis of the thousands of in vitro cytotoxicity, mutagenicity, or transformation studies, the objective of these points is that a careful re-analysis must be done on these studies before the results are used to draw risk assessment conclusions about the presumptive mechanisms by which either chemicals or radiations "cause" cancer.

WHAT CAN MOLECULAR ONCOLOGISTS SAY ABOUT WHAT CAUSES HUMAN CANCER AND CAN EPIDEMIOLOGISTS DETECT A HUMAN CARCINOGEN?

From the preceding discussion, it might seem that these rhetorical questions are similar to "How many angels can dance on a microchip"? However, while it is true that molecular oncologists are still a long way from understanding human carcinogenesis because the reductionalistic paradigms, and molecular and cellular assays have many limitations, we are not totally ignorant of factors influencing carcinogenesis. In addition, while it is also true that epidemiology is too insensitive to provide many answers, here again valuable insights have and will be gained from this field.

Recent reviews of human cancers associated with various kinds of radiation exposures have clearly shown that radiation can contribute to specific types of cancers.[180-183,241-243] There are very interesting observations made by these studies: (a) risk to radiation induced leukemias occurred when children were exposed to A-bomb radiation;[182,244] (b) certain solid tumors appeared later in life after exposure to A-bomb radiation;[245] (c) young females exposed to A-bomb radiation were at higher risk to breast cancer than older women;[246] (d) interaction between ionizing radiation and ultraviolet light exposure appears to exist in skin cancer;[247,248] (e) certain types of cancer give the superficial appearance of being nonradiogenic.[245]

In view of the previous analysis, it might be speculated that the stem cell and multistage theories of carcinogenesis are relevant to the breast cancer example. If the stem cells are the target cells for radiation, then the pregnancy status might be expected to affect the stem cell pool of the breast tissue. In other words, each pregnancy might be expected to remove stem cells from the pool by inducing terminal differentiation of the milk-producing cell types. The earlier the first pregnancy, and the more frequent the pregnancies before radiation would reduce the target size. In addition, this observation might be expected to imply the initiation/promotion/progression concept, since it takes many years from the acute radiation exposure before the appearance of the tumor. How radiation influenced this multistage process is not known. Did it initiate the process? If so, did it activate an oncogene or inactivate a tumor suppressor gene? Did it do so by a mutagenic event or by an epigenetic mechanism? If it initiated the stem cell, what are or could be endogenous or exogenous promoters? If older women are exposed to the same radiation and have accumulated many initiated cells during life, will they be at higher risk due to the predicted tumor suppressor gene deactivation or progressor potential of radiation?

There is no question that the time is here for more intense cross-fertilization of ideas to take place between these widely diverse disciplines.

SUMMARY

Cancers do appear in animals and human beings after exposure to ionizing radiation. However, the critical word in the title is "cause." In view of our current understanding of experimental in vitro and in vivo studies, as well as epidemiological data, carcinogenesis is the result of *many* external and endogenous factors. No single factor or "carcinogen" "causes" cancer. Carcinogenesis involves many steps, and several mechanisms for each step, with the interaction of external determinants, such as chemical and physical pollutants, medication/drugs, mutagenic and epigenetic agents—as these occur in the diet or workplace and environmental pollutants—and endogenous factors related to genetic background, sex, developmental stage, number of stem or progenital cells, growth factors, oncogenes, tumor suppressor and anti-metastasis genes. To generate risk assessments of potential cancer after exposures to radiations with only empirical epidemiological data is surely destined to fail without having a biologically-based model to account for the aforementioned factors.

A number of extant theories of carcinogenesis and of ionizing radiation's role in the process must be integrated. As a start in this process, an attempt will be made to integrate the "stem cell" theory, the theory of "oncogeny as partially blocked ontogeny," or the theory of cancer as a "disease of differentiation," the initiation/promotion/progression model of carcinogenesis, the oncogene/tumor suppressor gene theory and the mutation/epigenetic mechanisms of carcinogenesis. To do this, the phenomenon of intercellular communication (extra-, intra-, and inter-cellular) will be used as the integrating element, together with the current known facts of the types of mutations induced by ionizing radiation.

Given that radiation can potentially induce mutations, cell death, and epigenetic changes, the question arises as to how these biological consequences of ionizing radiation relate to these aforementioned theories. If ionizing radiation induces primarily deletion mutations and clastogen/chromosomal rearrangements, how does this relate to whether it is an initiator, promoter, or progressor of cancer? Does it affect oncogenes or tumor suppressor genes or both? Are all the cells of the body targets for the radiation's effects, or are there just special "target" cells? In what ways can genetic, developmental, and sex factors influence the radiation effects at each of the stages of carcinogenesis? In addition, how do exogenous factors influence the radiation effects on each of these stages?

Since not all genes nor all cells affected by radiation will lead to cancer, the question of threshold levels arises for each of the stages which radiation could affect. A number of factors bearing on this point are: (a) randomness of DNA damage within and between genes; (b) the existence of DNA repair mechanisms; (c) only a finite number of cancer related genes exist; (d) carcinogenesis needs more than one "hit," as it is more than mutagenesis; (e) kinetics of various types of induced mutations (e.g., initiation) will be different from induced mitogenesis of the initiated cell (e.g., promotion) and from induced conversion of initiated cell to malignant cell (e.g., progression).

In summary, while our understanding of radiation carcinogenesis is poorly understood today, we do know from the explosion of new findings and concepts of the multistep nature of carcinogenesis that our paradigms shaping current risk assessment models need dramatic revision. New emerging models should help shape new experimental approaches to derive a biologically-relevant risk assessment model.

ACKNOWLEDGMENTS

The authors wish to acknowledge the excellent technical assistance of Heather Rupp and Beth Lockwood for some of the research on which this paper was based, and the skilled word processing of Jeanne McHugh. The research on which this manuscript is based was supported, in part, by grants from the U.S. Air Force Office of Scientific Research (USAFOSR-F49626-92-J-0293), the NIEHS (1P42ESO4911) and the National Cancer Institute (CA21104).

REFERENCES

1. Pitot, H.C., T. Goldsworthy, and S. Moran, "The Natural History of Carcinogenesis: Implications of Experimental Carcinogenesis in the Genesis of Human Cancer," *J. Supramol. Struct. Cellul. Biochem.*, 17:133 (1981).
2. Pierce, G.B., "Neoplasms, Differentiation and Mutations," *Am. J. Pathol.*, 77:103 (1974).
3. Potter, V.R., "Phenotypic Diversity in Experimental Hepatomas: The Concept of Partially Blocked Ontogeny," *Br. J. Cancer*, 38:1 (1978).
4. Nowell, P.C., "The Clonal Evolution of Tumor Cell Population," *Science*, 194:23 (1976).
5. Fialkow, P.J., "Clonal Origin of Human Tumors," *Am. Rev. Med.*, 30:135 (1979).
6. Markert, C., "Neoplasia: A Disease of Cell Differentiation," *Cancer Research*, 28:1908 (1968).

7. Trosko, J.E., and C.C. Chang, "The Role of Radiation and Chemicals in the Induction of Mutations and Epigenetic Changes During Carcinogenesis," in *Advances in Radiation Biology*, J. Lett and H. Adler, Eds., (New York: Academic Press, 1981), p. 1.

8. Upton, A.C., "The Question of Thresholds for Radiation and Chemical Carcinogenesis," *Cancer Inv.*, 7:267 (1989).

9. Loewenstein, W.R., "Junctional Intercellular Communication: The Cell to Cell Membrane Channel," *Physiological Reviews*, 61:829 (1981).

10. Trosko, J.E., and C.C. Chang, "Implications for Risk Assessment of Genotoxic and Non-Genotoxic Mechanisms in Carcinogenesis," in *Methods for Estimating Risk of Chemical Injury: Human and Non-Human Biota Ecosystems*, V.B. Vouk, G.C. Butler, D.G. Hoel, and D.B. Peakall, Eds., (Chichester, England: John Wiley and Sons, 1985), p. 181.

11. Eagle, H., "Growth Regulatory Effects of Cellular Interaction," *Israel J. Med. Sci.*, 1:1220 (1965).

12. Brody, H., "A Systems View of Man: Implications for Medicine, Science and Ethics," *Perspect. Biol. Med.*, 17:71 (1973).

13. Potter, V.R., "Probabilistic Aspects of the Human Cybernetic Machine," *Perspect Biol. Med.*, 17:164 (1974).

14. Iverson, O.H., "Cybernetic Aspects of the Cancer Problem," in *Progress in Biocybernetics*, N. Wiener and J.P. Schade, Eds., (Amsterdam: Elsevier Publishing Company, 1965), p. 76.

15. Trosko, J.E., C.C. Chang, E. Dupont, B.V. Madhukar, and G. Kalimi, "Chemical Modulation of Gap Junctional Intercellular Communication in Vitro: An in Vitro Biomarker of Epigenetic Toxicology," in *In Vitro Methods in Toxicology*, G. Jolles and A. Cordier, Eds., (London: Academic Press, 1992), p. 465.

16. Trosko, J.E., "A Failed Paradigm: Carcinogenesis is More Than Mutagenesis, *Mutagenesis*, 3:363 (1988).

17. Trosko, J.E., "A New Paradigm Is Needed in Toxicological Evaluation," *Environ. Mutagen.*, 6:767 (1984).

18. Trosko, J.E., "Modulation of Gap Junction Function: The Scientific Basis of Epigenetic Toxicology," in *Environmental and Industrial Toxicology*, M. Ruchivawat, Ed., 1994, in press.

19. Ames, B.N., W.E. Durston, E. Yamasaki, and F.D. Lee, "Carcinogens Are Mutagens: A Simple Test System Combining Liver Homogenates for Activation and Bacteria for Detection," *Proc. Natl. Acad. Sci. USA*, 70:2281 (1973).

20. Trosko, J.E., "Critique of Intercellular Communication in Bronchial Epithelial Cells: A Review of Evidence for a Possible Role in Lung Carcinogenesis," *Toxicol. Pathology*, 18:341 (1990).

21. Trosko, J.E., C.C. Chang, B.V. Madhukar, and S.Y. Oh, "Modulators of Gap Junction Function: The Scientific Basis of Epigenetic Toxicology," *In Vitro Toxicol.*, 3:9 (1990).

22. Rodriguez-Tarduchy, G., and A. Lopez-Rivas, "Phorbol Esters Inhibit Apoptosis in IL-2-Dependent Lymphocytes-T," *Biochem. Biophys. Res. Commun.*, 164:1069 (1989).

23. Tomei, L.D., P. Kanter, and C.E. Wenner, "Inhibition of Radiation-Induced Apoptosis in Vitro by Tumor Promoters," *Biochem. Biophys. Res. Commun.*, 155:324 (1988).

24. Alles, A.J., and K.K. Sulik, "Retinoic-Acid-Induced Limb-Reduction Defects: Perturbation of Zones of Programmed Cell Death as a Pathogenetic Mechanism," *Teratology*, 40:163 (1989).

25. Garcea, R., L. Daino, R. Pascale, M.M. Simile, M. Puddu, S. Frassetto, P. Cozzolino, M.A. Seddaiu, L. Gaspa, and F. Feo, "Inhibition of Promotion and Persistent Nodule Growth by 5-Adenosyl-L-methionine in Rat Liver Carcinogenesis: Role of Remodeling and Apoptosis," *Cancer Res.*, 49:1850 (1989).

26. Bursch, W., B. Lauer, I. Timmermann-Trosiener, G. Barthel, J. Schuppler, and R. Schulte-Hermann, "Controlled Death (Apoptosis) of Normal and Putative Preneoplastic Cells in Rat Liver Following Withdrawal of Tumor Promoters," *Carcinogenesis*, 5:453 (1984).

27. Trosko, J.E., C.C. Chang, and A. Medcalf, "Mechanisms of Tumor Promotion: Potential Role of Intercellular Communication," *Cancer Invest.*, 1:511 (1983).

28. Farber, E., and H. Rubin, "Cellular Adaptation in the Origin and Development of Cancer," *Cancer Res.*, 51:2751 (1991).

29. Arguis, T.S., *CRC Reviews in Toxicology*, 14:211 (1985).

30. Frei, J.V., "Some Mechanisms Operative in Carcinogenesis: A Review," *Chem. Biol. Interact.*, 12:1 (1976).

31. Potter, V.R., "Use of Two Sequential Applications of Initiators in the Production of Hepatomas in the Rat: An Examination of the Solt-Farber Protocol," *Cancer Research*, 44:2733 (1984).

32. Trosko, J.E., and C.C. Chang, "An Integrative Hypothesis Linking Cancer, Diabetes and Atherosclerosis: The Role of Mutations and Epigenetic Changes," *Med. Hypoth.*, 6:455 (1980).

33. Trosko, J.E., and C.C. Chang, "Role of Intercellular Communication in Modifying the Consequences of Mutations in Somatic Cells," in *Antimutagenesis and Anticarcinogenesis Mechanisms*, D.M. Shankel, P.D. Hartman, T. Kada, and A. Hollaender, Eds., (New York: Plenum Press, 1986), p. 439.

34. Trosko, J.E., and C.C. Chang, "The Role of Inhibited Intercellular Communication in Carcinogenesis: Implications for Risk Assessment from Exposure to Chemicals," in *Biologically Based Methods for Cancer Risk Assessment*, C.C. Travis, Ed., (New York: Plenum Press, New York, 1989), p. 165.

35. Boutwell, R.K., "The Function and Mechanisms of Promoters of Carcinogenesis," *CRC Crit. Rev. Toxicol.*, 2:419 (1974).

36. O'Connell, J.F., A.J.P. Klein-Szanto, D.M. Digiovanni, J.M. Fries, and T.J. Slaga, "Malignant Progression of Mouse Skin Papillomas Treated with Ethylnitrosourea, NMNG, or TPA," *Cancer Lett.*, 30:269 (1986).

37. Reddy, A.L., and P.J. Fialkow, "Evidence That Weak Promotion of Carcinogen-Initiated Cells Prevents Their Progression to Malignancy," *Carcinogenesis*, 11:2123 (1990).

38. Hennings, H., R. Shores, M.L. Wenk, E.F. Spangler, R. Tarone, and S.H. Yuspa, "Malignant Conversion of Mouse Skin Tumors Is Increased by Tumor Initiators and Unaffected by Tumor Promoters," *Nature*, 304:67 (1983).

39. Pauli, B.U., and R.S. Weinstein, "Structure of Gap Junctions in Culture of Normal and Neoplastic Bladder Epithelial Cells," *Experientia*, 37:248 (1981).

40. Taguchi, T., M. Yokoyama, and Y. Kitamura, "Intraclonal Conversion from Papilloma to Carcinoma in the Skin of PgK-1^9/PgK-1^6 Mice Treated by a

Complete Carcinogenesis Process or by an Initiation-Promotion Regimen," *Cancer Res.*, 44:3779 (1984).

41. Jaffe, D.R., J.R. Williamson, and G.T. Bowden, "Ionizing Radiation Enhances Malignant Progression of Mouse Skin Tumors," *Carcinogenesis*, 8:1753 (1987).

42. Chu, K.C., C.C. Brown, R.E. Tarone, and W.Y. Tan, "Differentiating Among Proposed Mechanisms for Tumor Promotion in Mouse Skin with the Use of Multievent Model for Cancer," *J. Natl. Canc. Inst.*, 79:789 (1987).

43. Kerbel, R.S., P. Frost, R. Liteplo, D.A. Carlow, and B.E. Elliott, "Possible Epigenetic Mechanisms of Tumor Progression: Induction of High-Frequency Heritable But Phenotypically Unstable Changes in the Tumorigenic and Metastatic Properties of Tumor Cell Populations by 5-Azacytidine Treatment," *J. Cell. Physiol.*, 3:87 (1984).

44. Trosko, J.E., and C.C. Chang, "Stem Cell Theory of Carcinogenesis," *Toxicol. Lett.*, 49:283 (1989).

45. Trosko, J.E., C.C. Chang, and B.V. Madhukar, "Cell-to-Cell Communication: Relationship of Stem Cells to the Carcinogenic Process," *Prog. Clin. Biol. Res.*, 331: 259 (1990).

46. Nakano, S., H. Ueo, S.A. Bruce, and P.O.P. Ts'o, "A Contact-Insensitive Subpopulation in Syrian Hamster Cell Cultures with a Greater Susceptibility to Chemically Induced Neoplastic Transformation," *Proc. Natl. Acad. Sci. USA*, 82:5005 (1985).

47. Potten, C.S., Y.Q. Li, P.J. O'Connor, and D.J. Winton, "A Possible Explanation for the Differential Cancer Incidence in the Intestine, Based on the Distribution of the Cytotoxic Effects of Carcinogens in the Marine Large Bowel," *Carcinogenesis*, 13:2305 (1992).

48. Elmore, L.W., and A.E. Sirica, "Intestinal-Type of Adenocarcinoma Preferentially Induced in Right/Caudate Liver Lobes of Rats Treated with Furan," *Cancer Res.*, 57:254 (1993).

49. Slaga, T.J., R.K. Boutwell, and A. Sivak, *Carcinogenesis—A Comprehensive Survey, Vol. 2: Mechanisms of Tumor Promotion and Cocarcinogenesis*, (New York: Raven Press, 1978).

50. Miller, E.C., and J.A. Miller, "Mechanisms of Chemical Carcinogenesis," *Cancer Res.*, 47:1055 (1981).

51. Sisskin, E.E., T. Gray, and J.C. Barrett, "Correlation Between Sensitivity to Tumor Promotion and Sustained Epidermal Hyperplasia of Mice and Rats," *Carcinogenesis*, 3:403 (1982).

52. Boveri, T., *zur Frage der Entstehung Maligner Tumoren*, (Jena: Gustav Fisher, Pub., 1914).

53. Trosko, J.E., V.M. Riccardi, C.C. Chang, S. Warren, and M. Wade, "Genetic Predisposition to Initiation or Promotion Phases in Human Carcinogenesis," in *Biomarkers, Genetics and Cancer*, H. Anton-Guirgis and H.T. Lynch, Eds., (New York: Van Nostrand Reinhold Company, 1985), p. 13.

54. Trosko, J.E., "Cancer Causation," *Nature*, 290:356 (1981).

55. Maher, V.M., and J.J. McCormick, "Effect of DNA Repair on the Cytotoxicity and Mutagenicity of UV Irradiation and of Chemical Carcinogens in Normal and Xeroderma Pigmentosum Cells," in *Biology of Radiation Carcinogenesis*, J.M. Yuhas, R.W. Tennant, and J.D. Regan, Eds., (New York: Raven Press, 1976), p. 129.

56. Glover, T.W., C.C. Chang, J.E. Trosko, and S.S.L. Li, "Ultraviolet Light Induction of Diphtheria Toxin Resistant-Mutants in Normal and Xeroderma Pigmentosum Human Fibroblasts," *Proc. Natl. Acad. Sci. USA*, 76:3982 (1979).
57. Warren, S.T., R.A. Schultz, C.C. Chang, M.H. Wade, and J.E. Trosko, "Elevated Spontaneous Mutation Rate in Bloom Syndrome Fibroblasts," *Proc. Natl. Acad. Sci. USA*, 78:3133 (1981).
58. Trosko, J.E., and C.C. Chang, "Nongenotoxic Mechanisms in Carcinogenesis: Role of Inhibited Intercellular Communication," in *Banbury Report 31: Carcinogen Risk Assessment: New Directions in the Qualitative and Quantitative Aspects*, R.W. Hart, and F.G. Hoerger, Eds., (Cold Spring Harbor, NY: Cold Spring Harbor Laboratory Press, 1988), p. 139.
59. Pitot, H.C., "Endogenous Carcinogenesis: The Role of Tumor Promotion," *Proceedings of the Society for Experimental Biology and Medicine*, 198:661 (1991).
60. Yager, J.D., and R. Yager, "Oral Contraceptive Steroids as Promoters of Hepatocarcinogenesis in Female Sprague-Dawley Rats," *Cancer Research*, 40:3680 (1980).
61. Rose, S.P., R. Stahn, D.S. Passovoy, and H. Herschman, "Epidermal Growth Factor Enhancement of Skin Tumor Induction in Mice," *Experientia*, 32:913 (1976).
62. Harrison, J., and N. Auersperg, "Epidermal Growth Factor Enhances Viral Transformation of Granulosa Cells," *Science*, 213:218 (1981).
63. Lipton, A., N. Kepner, C. Rogers, E. Witkoski, and K. Leitzel, "A Mitogenic Factor for Transformed Cells from Human Platelets, in *Interactions of Platelets and Tumor Cells*, G.A. Jamieson, Ed., (New York: Alan R. Liss, Inc., 1982), p. 233.
64. Stoscheck, C.M., and L.E. King, "Role of Epidermal Growth Factor in Carcinogenesis," *Cancer Res.*, 46:1030 (1986).
65. Chester, J.F., H.A. Gaissert, J.S. Ross, and R.A. Malt, "Pancreatic Cancer in the Syrian Hamster Induced by N-Nitrosobis(2-oxopropyl)-amine Cocarcinogenic Effect of Epidermal Growth Factor," *Cancer Res.*, 46:2954 (1986).
66. Fisher, P.B., R.A. Mufson, I.B. Weinstein, and J.B. Little, "Epidermal Growth Factor, Like Tumor Promoters, Enhances Viral and Radiation-Induced Cell Transformation," *Carcinogenesis*, 2:183 (1981).
67. Mordan, L.J., "Induction by Growth Factors from Platelets of the Focus-Forming Transformed Phenotype in Carcinogen-Treated C3H/10T1/2 Fibroblasts," *Carcinogenesis*, 9:1129 (1988).
68. Lee, W.H., R. Bookstein, F. Hong, L.J. Young, J.Y. Shew, and E.Y. Lee, "Human Retinoblastoma Susceptibility Gene: Cloning, Identification and Sequence," *Science*, 235:1394 (1987).
69. Malkin, D., F.P. Li, L.C. Strong, J.F. Fraumeni, C.E. Nelson, D.H. Kim, J. Kassel, M.A. Gryka, F.Z. Bischoff, M.A. Tainsky, and S.H. Friend, "Germ Line p53 Mutations in a Familial Syndrome of Breast Cancer Sarcomas, and Other Neoplasms," *Science*, 250:1233 (1990).
70. Weinberg, R.A., "The Action of Oncogenes in the Cytoplasm and Nucleus," *Science*, 230:770 (1985).
71. Bishop, J.M., "Viral Oncogenes," *Cell*, 42:23 (1985).
72. Weinberg, R.A., "Tumor Suppressor Genes," *Science*, 254:1138 (1991).

73. Lacey, S.W., "Oncogenes in Retroviruses, Malignancy and Normal Tissues," *Amer. J. Med. Sci.*, 29:39 (1986).
74. Newbold, D.F., and R.W. Overell, "Fibroblast Immortality is a Prerequisite for Transformation by EJ c-Has-ras Oncogene," *Nature*, 304:648 (1983).
75. Land, H., L.F. Parada, and R.A. Weinberg, "Tumorigenic Conversion of Primary Embryo Fibroblasts Requires at Least Two Cooperating Oncogenes," *Nature*, 304:596 (1983).
76. Spandidos, D.A., and M.L.M. Anderson, "A Study of Mechanisms of Carcinogenesis by Gene Transfer of Oncogenes into Mammalian Cells, *Mutat. Res.*, 185:271 (1987).
77. Stanbridge, E.J., "Identifying Tumor Suppressor Genes in Human Colorectal Cancer," *Science*, 247:12 (1990).
78. Vogelstein, B., E.R. Fearon, R. White, A.M.M. Smits, and J.L. Bos, "Genetic Alterations During Colorectal Tumor Development," *New Engl. J. Med.*, 319:525 (1988).
79. Annab, L.A., J.T. Dong, A. Futreal, H. Satoh, M. Oshimura, and J.C. Barrett, "Growth and Transformation Suppressor Genes for BHK Syrian Hamster Cells on Human Chromosome 1 and 11," *Molec. Carcinogenesis*, 6:280.
80. Larsen, W.J., and M.A. Risinger, "The Dynamic Life Histories of Intercellular Membrane Junctions," *Mod. Cell Biol.*, 4:151 (1985).
81. Madhukar, B.V., S.Y. Oh, C.C. Chang, M. Wade, and J.E. Trosko, "Altered Regulation of Intercellular Communication by Epidermal Growth Factor, Transforming Growth Factor-Beta and Peptide Hormones in Normal Human Keratinocytes," *Carcinogenesis*, 10:13 (1989).
82. Maldonado, P.E., B. Rose, and W.R. Loewenstein, "Growth Factors Modulate Junctional Cell-to-Cell Communication," *J. Memb. Biol.*, 106:203 (1988).
83. Neyton, J., and A. Trautman, "Acetylcholine Modulation of the Conductance of Intercellular Junctions Between Rat Lacrimal Cells," *J. Physiol.*, 377:283 (1986).
84. Revel, J.P., "The Oldest Multicellular Animal and Its Junctions," in *Gap Junctions*, E.L. Hertzberg and R.G. Johnston, Eds., (New York: Alan R. Liss, Inc., 1988), p. 135.
85. Loewenstein, W.R., "Junctional Intercellular Communication and the Control of Growth," *Biochim. Biophys. Acta*, 560:1 (1979).
86. Lo, C.W., "Communication Compartmentation and Pattern Formation in Development," in *Gap Junctions*, M.V.L. Bennett and D.C. Spray, Eds., (Cold Springs Harbor, NY: Cold Spring Harbor Laboratory, 1985), p. 251.
87. Meda, P., R. Bruzzone, M. Chanson, D. Bosco, and L. Orci, "Gap Junctional Coupling Modulates Secretion of Exocrine Pancreas," *Proc. Natl. Acad. Sci. USA*, 84:4901 (1987).
88. Vesch, F., A. Schafer, and R.J. Wieser, "12-0-Tetradecanoylphorbol-13-acetate Releases Human Diploid Fibroblasts from Contact-Dependent Inhibition of Growth," *Carcinogenesis*, 9:1319 (1988).
89. DeMello, W.C., "Cell-to-Cell Communication in Heart and Other Tissues," *Prog. Biophys. Mol. Biol.*, 39:147 (1982).
90. Cole, W.C., and R.E. Garfield, "Evidence for Physiological Regulation of Myometrial Gap Junction Permeability," *Am. J. Physiol.*, 251:411 (1986).
91. Sheridan, J.D., "Cell Communication and Growth," in *Cell-to-Cell Communication*, W.C. DeMello, Ed., (New York: Plenum Press, 1987), p. 187.

92. Spray, D.C., and M.V.L. Bennett, "Physiology and Pharmacology of Gap Junctions," *Ann. Rev. Physiol.*, 47:281 (1985).

93. Saez, J.C., D.C. Spray, and E.L. Hertzberg, "Gap Junctions: Biochemical Properties and Functional Regulation under Physiological and Toxicological Conditions," *In Vitro Toxicol.*, 3:69 (1990).

94. Saez, J.C., M.V.L. Bennett, and D.C. Spray, "Carbon Tetrachloride at Hepatotoxic Levels Blocks Reversibly Gap Junctions Between Rat Hepatocytes," *Science*, 236:967 (1987).

95. Yancey, S.B., D. Easter, and J.P. Revel, "Cytological Changes in Gap Junctions During Liver Regeneration," *J. Ultrastruct. Res.*, 67:229 (1979).

96. Traub, O., P.M. Druge, and K. Willecke, "Degradation and Resynthesis of Gap Junction Protein in Plasma Membranes of Regenerating Liver after Partial Hepatectomy or Cholestasis," *Proc. Natl. Acad. Sci. USA*, 80:755 (1983).

97. Borek, C., and L. Sachs, "The Difference in Contact Inhibition of Cell Replication Between Normal Cells and Cells Transformed by Different Carcinogens," *Proc. Natl. Acad. Sci. USA*, 56:1705 (1966).

98. Kanno, Y., "Modulation of Cell Communication and Carcinogenesis," *Jpn. J. Physiol.*, 35:693 (1985).

99. Fentman, I.S., J. Hurst, R.L. Ceriani, and J. Taylor-Papadimitriou, "Junctional Intercellular Communication Pattern of Cultured Human Breast Cancer Cells," *Cancer Res.*, 39:4739 (1979).

100. Albright, C.D., P.M. Grimley, R.T. Jones, J.A. Fontana, K.P. Keenan, and J.H. Resau, "Cell-to-Cell Communication: A Differential Response to TGF-beta in Normal and Transformed (BEAS-2B) Human Bronchial Epithelial Cells," *Carcinogenesis*, 12:1993 (1991).

101. Smith, B.R., "Regulation of Hematopoiesis," *Yale J. Biol. Med.*, 63:371 (1990).

102. Dainiak, N., "Surface Membrane-Associated Regulation of Cell Assembly, Differentiation, and Growth, *Blood*, 78:264 (1991).

103. Rosendaal, M., A. Gregan, and C.R. Green, "Direct Cell-Cell Communication in the Blood-Forming System," *Tissue Cell*, 23:457 (1991).

104. Chang, C.C., J.E. Trosko, M.H. El-Fouly, R.E. Gibson-D'Ambrosio, and S.M. D'Ambrosio, "Contact Insensitivity of a Subpopulation of Normal Human Fetal Kidney Epithelial Cells and of Human Carcinoma Cell Lines," *Cancer Res.*, 47:1634 (1987).

105. Chang, C.C., S. Nakatsuka, G. Kalimi, J.E. Trosko, and C.W. Welsch, "Characterization of Two Types of Normal Human Breast Epithelial Cells That Are Either Deficient or Proficient in Gap Junctional Intercellular Communication," *J. Cell Biochem. Suppl.*, 14B:331 (1990).

106. Keski, J., and H.C. Moses, "Growth Inhibitory Polypeptide in the Regulation of Cell Proliferation," *Med. Biol.*, 65:13 (1987).

107. Sonnenschein, C., N. Olea, M.E. Pasanen, and A.M. Soto, "Negative Controls of Cell Proliferation: Human Prostate Cancer Cells and Androgens," *Cancer Res.*, 49:3474 (1989).

108. Kimchi, A., X.-F. Wang, R.A. Weinberg, S. Cheifetz, and J. Massague, "Absence of TGF-beta Receptors and Growth Inhibitory Responses in Retinoblastoma Cells," *Science*, 240:196 (1988).

109. de Rooij, D.G., D. Lok, and D. Weenk, "Feedback Regulation of the Prolifer-

ation of the Undifferentiated Spermatogonia in the Chinese Hamster by the Differentiating Spermatogonia," *Cell Tissue Kinet.*, 18:71 (1985).

110. Trosko, J.E., C.C. Chang, B.V. Madhukar, and E. Dupont, "Oncogenes, Tumor Suppressor Genes and Intercellular Communication in the 'Oncogeny as Partially Block Ontogeny' Hypothesis," *Theories of Carcinogenesis*, in press.

111. Weinstein, R.S., F.B. Merk, and J. Alroy, "The Structure and Function of Intercellular Junctions in Cancer," *Adv. Cancer Res.*, 23:23 (1976).

112. Yamasaki, H., M. Hollstein, M. Mesnil, N. Martel, and A.M. Aguelon, "Selective Lack of Intercellular Communication Between Transformed and Nontransformed Cells as a Common Property of Chemical and Oncogene Transformation of BALB/c3T3 Cells," *Cancer Res.*, 47:5658 (1987).

113. Enomoto, T., and H. Yamasaki, "Lack of Intercellular Communication Between Chemically Transformed and Surrounding Nontransformed BALB/c3T3 Cells," *Cancer Res.*, 44:5200 (1984).

114. Sugie, S., H. Mori, and M. Takahashi, "Effect of in Vivo Exposure to the Liver Tumor Promoters Phenobarbital or DDT on the Gap Junctions of Rat Hepatocytes: A Quantitative Freeze-Fracture Analysis," *Carcinogenesis*, 8:45 (1987).

115. Janssen-Timmen, U., O. Traub, R. Dermietzel, H.M. Rabes, and K. Willecke, "Reduced Number of Gap Junctions in Rat Hepatocarcinomas Detected by Monoclonal Antibody," *Carcinogenesis*, 7:1475 (1986).

116. Swift, J.G., T.M. Mukherjee, and R. Rowland, "Intercellular Junctions in Hepatocellular Carcinoma," *J. Submicrosc. Cytol. Pathol.*, 15:799 (1983).

117. Loewenstein, W.R., and Y. Kanno, "Intercellular Communication and the Control of Tissue Growth: Lack of Communication Between Cancer Cells," *Nature*, 209:1248 (1966).

118. Kalimi, G.H., and S.M. Sirsat, "The Relevance of Gap Junctions to Stage I Tumor Promotion in Mouse Epidermis," *Carcinogenesis*, 5:1671 (1984).

119. Lincoln, D.W., S.J. Kampcik, and J.F. Gierthy, 2,3,7,8-Tetrachlorodibenzo-p-dioxin (TCDD) Does Not Inhibit Intercellular Communication in Chinese Hamster V79 Cells," *Carcinogenesis*, 8:1817 (1987).

120. Atkinson, M.M., A.S. Menko, R.G. Johnson, I.R. Sheppard, and J.D. Sheridan, "Rapid and Reversible Reduction of Junctional Permeability in Cells Infected with a Temperature Sensitive Mutant of Avian Sarcoma Virus," *J. Cell Biol.*, 9:573 (1981).

121. Atkinson, M.M., S.K. Anderson, and J.D. Sheridan, "Modification of Gap Junctions in Cells Transformed by a Temperature-Sensitive Mutant of Rous Sarcoma Virus," *J. Membr. Biol.*, 91:53 (1986).

122. Azarnia, R., and W.R. Loewenstein, "Intercellular Communication and the Control of Growth: Alteration of Junctional Permeability by the src Gene — A Study with Temperature-Sensitive Mutant Rous Sarcoma Virus," *J. Membr. Biol.*, 82:191 (1984).

123. Azarnia, R., S. Reddy, T.E. Kimiecki, D. Shalloway, and W.R. Loewenstein, "The Cellular src Gene Product Regulates Junctional Cell-to-Cell Communication," *Science*, 239:398 (1988).

124. Azarnia, R., and W.R. Loewenstein, "Polyomavirus Middle T Antigen Downregulates Junctional Cell to Cell Communication," *Mol. Cell Biol.*, 7:946 (1987).

125. Chang, C.C., J.E. Trosko, H.J. Kung, D. Bombick, and F. Matsumura, "Potential Role of the src Gene Product in Inhibition of Gap Junctional Communication in NIH 3T3 Cells," *Proc. Natl. Acad. Sci. USA*, 82:5360 (1985).

126. Dotto, G.P., M.H. El-Fouly, C. Nelson, and J.E. Trosko, "Similar and Synergistic Inhibition of Gap-Junctional Communication by ras Transformation and Tumor Promoter Treatment of Mouse Primary Keratinocytes," *Oncogene*, 4:637 (1989).

127. Bignami, M., S. Rosa, G. Falcone, F. Tato, F. Katoh, and H. Yamasaki, "Specific Viral Oncogenes Cause Differential Effects on Cell-to-Cell Communication Relevant to the Suppression of the Transformed Phenotype by Normal Cells," *Molec. Carcinogenesis*, 1:67 (1988).

128. El-Fouly, M.H., J.E. Trosko, and C.C. Chang, "Phenotypic Transformation and Inhibition of Gap-Junctional Intercellular Communication in Epithelial and Mesenchymal Cells by the Neu Oncogene," *4th Annual Oncogene Meeting*, Frederick, MD, July 5–9, 1988, abstract.

129. El-Fouly, M.H., J.E. Trosko, C.C. Chang, and S.T. Warren, "Potential Role of the Human Ha-ras Oncogene in the Inhibition of Gap Junctional Intercellular Communication," *Mol. Carcinog.*, 2:131 (1989).

130. de Feijter, A.W., J.S. Ray, C.M. Weghorst, J.E. Klaunig, J.I. Goodman, C.C. Chang, R.J. Ruch, and J.E. Trosko, "Infection of Rat Liver Epithelial Cells with v-Ha-ras: Correlation Between Oncogene Expression, Gap Junctional Communication, and Tumorigenicity," *Mol. Carcinog.*, 3:54 (1990).

131. Chang, J.M., and G.R. Johnson, "Gene Transfer into Hemopoietic Stem Cells Using Retroviral Vectors, "*Int. J. Cell*, 7:264 (1989).

132. Kikkawa, U., A. Kishimoto, and Y. Nishizuka, "The Protein Kinase-C Family—Heterogeneity and Its Implications, "*Ann. Rev. Biochem.*, 58:31 (1989).

133. Kalimi, G.H., L.L. Hampton, J.E. Trosko, S.S. Thorgeirsson, and A.C. Huggett, "Homologous and Heterologous Gap-Junctional Intercellular Communication in v-raf-, v-myc-, and v-raf/v-myc-Transducted Rat Liver Epithelial Cell Lines," *Mol. Carcinog.*, 5:301 (1992).

134. Lee, S.W., C. Tomasetto, and R. Sager, "Positive Selection of Candidate Tumor-Suppressor Genes by Subtractive Hybridization," *Proc. Natl. Acad. Sci. USA*, 88:2825 (1991).

135. Kalimi, G., C.C. Chang, P. Edwards, E. Dupont, B.V. Madhukar, E. Stanbridge, J.E. Trosko, "Re-Establishment of Gap Junctional Communication in a Non-Tumorigenic Hela-Normal Human Fibroblast Hybrid," *Proc. Am. Assoc. Cancer Res.*, 31:319 (1990).

136. Fujiki, H., "Is the Inhibition of Protein Phosphatase 1 and 2A Activities a General Mechanism of Tumor Promotion in Human Cancer Development?" *Molec. Carcinogenesis*, 5:91 (1992).

137. Rivedal, E., and R. Sanner, "Regulation of Gap Junctional Communication in Syrian Hamster Embryo Cells by Retinoic Acid and 12-O-Tetradecanoylphorbol-13-acetate," *Carcinogenesis*, 13:199 (1992).

138. Mehta, P.P., J.S. Bertram, and W.R. Loewenstein, "The Actions of Retinoids on Cellular Growth Correlate with Their Actions on Gap Junctional Communication," *J. Cell Biol.*, 108:1053 (1989).

139. Mehta, P.P., and W.R. Loewenstein, "Differential Regulation of Communica-

tion by Retinoic Acid in Homologous and Heterologous Junctions Between Normal and Transformed Cells," *Journal of Cell Biology*, 113:371 (1991).

140. Rogers, M., J.M. Berestecky, M.Z. Hossain, H.M. Guo, R. Kadle, B.J. Nicholson, and J.S. Bertram, "Retinoid-Enhanced Gap Junctional Communication Is Achieved by Increased Levels of Connexin-43 Messenger RNA and Protein," *Mol. Carcinogen.*, 3:335 (1990).

141. Brummer, F., G. Zempel, P. Buhle, J.C. Stein, and D.F. Hulser, "Retinoic acid Modulates Gap Junctional Permeability: A Comparative Study of Dye Spreading and Ionic Coupling in Cultured Cells," *Exp. Cell Res.*, 196:158 (1991).

142. Griesinger, F., B. Jansen, and J.H. Kersey, "Differentiation in Mature T-Lymphoid Leukemia Cells Is Unstable and Reversible to Myeloid Cells, Without the Involvement of a Common Stem Cell," *J. Immunol.*, 147:3336 (1991).

143. Azarnia, R., and T.R. Russell, "Cyclic AMP Effects on Cell to Cell Junctional Membrane Permeability During Adipocyte Differentiation of 3T3-L1 Fibroblasts," *J. Cell Biol.*, 100:265 (1985).

144. Flagg-Newton, J.L., G. Dahl, and W.R. Loewenstein, "Cell Junction and Cyclic AMP: Upregulation of Junctional Membrane Permeability and Junctional Membrane Particles by Administration of Cyclic Nucleotide or Phosphodiesterase Inhibitor," *J. Memb. Biol.*, 63:105 (1981).

145. Mehta, P.P., J.S. Bertram, and W.R. Loewenstein, "Growth Inhibition of Transformed Cells Correlates with the Junctional Communication with Normal Cells," *Cell*, 44:187 (1986).

146. Demaziere, A.M.G.L., and D.W. Scheuerman, "Increased Gap Junctional Area in the Rat Liver After Administration of Dibutyrl cAMP," *Cell Tiss. Res.*, 239:651 (1985).

147. Veld, P.I., F. Schuit, and D. Pipeleers, "Gap Junctions Between Pancreatic B-Cells Are Modulated by Cyclic AMP," *Eur. J. Cell Biol.*, 36:269 (1985).

148. Ruch, R.J., S.J. Cheng, and J.E. Klaunig, "Prevention of Cytotoxicity and Inhibition of Intercellular Communication by Antioxidant Catechins Isolated from Chinese Green Tea," *Carcinogenesis*, 10:1003 (1989).

149. Ruch, R.J., B.V. Madhukar, J.E. Trosko, and J.E. Klaunig, "Reversal of ras-Induced Inhibition of Gap Junctional Intercellular Communication, Transformation and Tumorigenesis by Lovastatin," *Molec. Carcinogenesis*, 7:50 (1993).

150. Casey, P.J., P.A. Solski, C.J. Der, and J.E. Buss, "p21ras is Modified by a Farnesylisoprenoid," *Proc. Natl. Acad. Sci. USA*, 86:8323 (1989).

151. Jackson, J.H., C.G. Cochrane, J.R. Bourne, P.A. Solski, J.E. Buss, and C.J. Der, "Farnesol Modification of Kirsten-ras exon 4B Protein Is Essential for Transformation," *Proc. Natl. Acad. Sci. USA*, 87:3042 (1990).

152. Sebti, S.M., G.T. Tkalcevic, and J.P. Jani, "Lovastatin, a Cholesterol Biosynthesis Inhibitor, Inhibits the Growth of Human H-ras Oncogene Transformed Cells in Nude Mice," *Cancer Commun.*, 3:141 (1991).

153. Fishman, G.I., A.P. Moreno, D.C. Spray, and L.A. Leinwand, "Functional Analysis of Human Cardiac Gap Junction Channel Mutants," *Proc. Natl. Acad. Sci. USA*, 88:3525 (1991).

154. Zhu, D., S. Caveney, G.M. Kidder, and C.C.G. Naus, "Transfection of C6 Glioma Cells with Connexin-43 cDNA: Analysis of Expression, Intercellular Coupling, and Cell Proliferation," *Proc. Natl. Acad. Sci. USA*, 88:1883 (1991).

155. Eghbali, B., J.A. Kessler, and D.C. Spray, "Expression of Gap Junction Channels in Communication-Incompetent Cells After Stable Transfection with cDNA Encoding Connexin-32," *Proc. National Academy of Science, USA*, 87:1328 (1990).

156. Zhu, D.G., G.M. Kidder, S. Caveney, and C.C.G. Naus, "Growth Retardation in Glioma Cells Cocultured with Cells Overexpressing a Gap Junction Protein," *Proc. Natl. Acad. Sci. USA*, 89:10218 (1992).

157. Oh, S.Y., E. Dupont, B.V. Madhukar, J.P. Briand, C.C. Chang, E. Beyer, and J.E. Trosko, "Characterization of Gap Junctional Communication Deficient Mutants of a Rat Liver Epithelial Cell Line, WB-F344," *Eur. J. Cell Biol.*, 60:250 (1992).

158. Muller, H.J., "Artificial Transmutation of the Gene," *Science*, 66:84 (1927).

159. Lieberman, M.W., R.N. Baney, R.E. Lee, S. Sell, and E. Farber, "Studies on DNA Repair in Human Lymphocytes Treated with Ultimate Carcinogens and Alkylating Agents," *Cancer Res.*, 3:1292 (1971).

160. Slaga, T.J., S.M. Fischer, C.E. Weeks, A. Klein-Szanto, and J. Reiners, "Studies on the Mechanisms Involved in Multistage Carcinogenesis in Mouse Skin," *J. Cell. Biochem.*, 18:99 (1982).

161. Kraemer, K.H., "Oculo-Cutaneous and Internal Neoplasms in Xeroderma Pigmentosum: Implications for Theories of Carcinogenesis," in *Carcinogenesis: Fundamental Mechanisms and Environmental Effects*, B. Pullman, P.O.P. Ts'o, and H. Gelboin, Eds., (New York: D. Reidel Publishing Co., 1980), p. 503.

162. Robbins, J.H., K.H. Kraemer, M.A. Lutzner, B.W. Festoff, and H.G. Coon, "Xeroderma Pigmentosum: An Inherited Disease with Sun Sensitivity, Multiple Cutaneous, Neoplasms and Abnormal DNA Repair," *Ann. Intern Med.*, 80:221 (1974).

163. Ashby, J., and R.S. Morrod, "Detection of Human Carcinogens," *Nature*, 352:185 (1991).

164. Mass, M.J., and S.J. Austin, "Absence of Mutations in Codon 61 of the Ha-ras Oncogene in Epithelial Cells Transformed in Vitro by 7,12-Dimethylbenz(a)-athracene," *Biochem. Biophys. Res. Commun.*, 165:1319 (1989).

165. Brookes, P., "Chemical Carcinogens and ras Gene Activation," *Molec. Carcinogenesis*, 2:305 (1989).

166. Mass, M.J., N.S. Schorschinsky, J.A. Lasley, D.K. Beeman, and S.J. Austin, "Oncogene Alterations in *in Vitro* Transformed Rat Tracheal Epithelial Cells," *Mutation Research*, 243:291 (1990).

167. Reynolds, S.H., S.J. Stowers, R.M. Patterson, R.R. Maronpot, S.A. Aaronson, and M.W. Anderson, "Activated Oncogenes in B6C3F-1 Mouse Liver Tumors: Implications for Risk Assessment," *Science*, 237:1309 (1987).

168. Stanley, L.A., T.R. Devereux, J. Foley, P.G. Lord, R.R. Maronpot, T.C. Orton, and M.W. Anderson, "Proto-Oncogene Activation in Liver Tumors of Hepatocarcinogenesis-Resistant Strains of Mice," *Carcinogenesis*, 13:2427 (1992).

169. Masui, T., A.M. Mann, T.L. Macatee, T. Okamura, E.M. Garland, R.A. Smith, and S.M. Cohen, "Absence of ras Oncogene Activation in Rat Urinary Bladder Carcinomas Induced by N-methyl-N-nitrosouvea or N-butyl-N-(4-hydroxybutyl) nitrosamine," *Carcinogenesis*, 13:2281 (1992).

170. Mann, A.M., M. Asamoto, F.J. Chlapowsk, T. Masui, T.L. Macatee, and S.M. Cohen, "Ras Involvement in Cells Transformed with 2-Amino-4-(5-nitro-2-furyl)thiazole (ANFT) in Vitro and with N-[4-(5-Nitro-2-furyl)-2-thiazolyl]formamide in Vivo," *Carcinogenesis*, 13:1651 (1992).

171. Chen, A.C., D.W. Brankow, and H.R. Herschman, "A Re-Assessment of Methylcholanthrene Transformation in the C3H10T1/2 Cell Culture System," *Carcinogenesis*, 11:817 (1990).

172. Fox, T.R., A.M. Schumann, P.G. Watanabe, B.L. Yano, V.M. Maher, and J.J. McCormick, "Mutational Analysis of the H-ras Oncogene in Spontaneous C57BL/6xC3H/He Mouse Liver Tumors and Tumors Induced with Genotoxic and Nongenotoxic Hepatocarcinogens," *Cancer Res.*, 50:4014 (1990).

173. Zhang, R., J.D. Haag, and M.N. Gould, "Reduction in the Frequency of Activated ras Oncogenes in Rat Mammary Carcinomas with Increasing N-Methyl-N-nitrosourea Doses on Increasing Prolactin Levels," *Cancer Res.*, 50:4286 (1990).

174. Schulte-Hermann, R., I. Timmermann-Trosiener, and J. Schuppler, "Promotion of Spontaneous Preneoplastic Cells in Rat Liver as a Possible Explanation of Tumor Production by Nonmutagenic Compounds," *Cancer Res.*, 43:839 (1983).

175. Ashby, J., and R.W. Tennant, "Chemical Structure, Salmonella Mutagenicity and Extent of Carcinogenicity as Indicators of Genotoxic Carcinogenesis Among 222 Chemicals Tested in Rodents by the U.S. NCI/NTP," *Mutat. Res.*, 204:17 (1988).

176. Trosko, J.E., C.C. Chang, B.V. Madhukar, and E. Dupont, "Intercellular Communication: A Paradigm for the Interpretation of the Initiation/Promotion/Progression Model of Carcinogenesis," in *Chemical Induction of Cancer: Modulation and Combination Effects*, J.C. Arcos, Ed., (New York: Academic Press, in press).

177. Fry, R.J.M., R.D. Rey, D. Grube, and E. Staffeldt, "Studies on the Multistage Nature of Radiation Carcinogenesis," in *Carcinogenesis*, H. Hecker, N.E. Fusenig, W. Kunz, F. Marks, and H.N. Thielmann, Eds., (New York: Raven Press, 1982), p. 155.

178. Fry, R.J.M., "Radiation Carcinogenesis in the Whole-Body System," *Radiat. Res.*, 126:157 (1991).

179. Elking, M.M., J.S. Bedford, S.A. Benjamin, E.A. Hoover, W.K. Sinclair, S.S. Wallace, and J.D. Zinbrick, "Oncogenic Mechanisms in Radiation-Induced Cancer," *Cancer Res.*, 51:2740 (1991).

180. Boice, J.D., "Carcinogenesis: A Synopsis of Human Experience with External Exposure in Medicine," *Health Phys.*, 55:621 (1988).

181. Beebe, G.W., "Ionizing Radiation and Health," *Amer. Sci.*, 70:35 (1982).

182. Shimizu, Y., W.J. Schull, and H. Kato, "Cancer Risk Among Atomic Bomb Survivors," *J. A. M. A.*, 264:601 (1990).

183. "The Future of Human Radiation Research," in *Brit. Instit. Radiol. Report 22*, G.B. Gerber, D.M. Taylor, E. Cardis, and J.M. Thiessen, Eds., London, 1991.

184. Borek, C., and L. Sachs, "In Vitro Cell Transformation by x-Irradiation," *Nature*, 16:276 (1966).

185. Watanabe, M., M. Horikawa, and O. Nikaido, "Induction of Oncogenic Transformation by Low Doses of x-Rays and Dose Rate Effect," *Radiat. Res.*, 98:274 (1984).

186. Han, A., and M.M. Elkind, "Transformation of Mouse C3H 10T1/2 Cells by Single and Fractionated Doses of x-Rays and Fission-Spectrum Neutrons," *Cancer Res.*, 39:123 (1979).

187. Trosko, J.E., "Does Radiation Cause Cancer?" in *RERF Update*, 1992, 3.

188. Fuscoe, J.C., C.H. Ockey, and M. Fox, "Molecular Analysis of x-Ray Induced Mutants at the HPRT Locus in V79 Chinese Hamster Cells," *Int. J. Radiat. Biol.*, 49:1011 (1986).

189. Liber, H.L., K.M. Call, and J.B. Little, "Molecular and Biochemical Analyses of Spontaneous and x-Ray Induced Mutants in Human Lymphoblastoid Cells," *Mutat. Res.*, 178:143 (1987).

190. Thacker, J., "The Nature of Mutants Induced by Ionizing Radiation in Cultured Hamster Cells. III. Molecular Characterization of HPRT Deficient Mutants Induced by Gamma-Rays or Alpha-Particles Showing That the Majority Have Deletions of All or Part of the HPRT Gene," *Mutat. Res.*, 160:267 (1986).

191. Fuscoe, J.C., L.J. Zimmerman, A. Fekete, R.W. Setzer, and B.J.F. Rossiter, "Analysis of X-Ray-Induced HPRT Mutations in CHO Cells: Insertion and Deletions," *Mutat. Res.*, 269:171 (1992).

192. Whaley, J.M., and J.B. Little, "Molecular Characterization of hprt Mutants Induced by Low- and High-LET Radiations in Human Cells," *Mutat. Res.*, 243:35 (1990).

193. Skandalis, A., A.J. Grosovsky, E.A. Drobetsky, and B.W. Glickman, "Investigation of the Mutagenic Specificity of X-Rays Using a Retroviral Shuttle Vector in CHO Cells," *Environ. Mol. Mutagen.*, 20:271 (1992).

194. Jou, Y.S., C.C. Chang, and J.E. Trosko, "Development of a Human Epithelial Cell Line with Increased Recovery of Point and Deletion Mutations," 1994, in press.

195. Neel, J.V., W.J. Schull, A.A. Awa, C. Satoh, H. Kato, M. Otake, and Y. Yoshimoto, "The Children of Parents Exposed to Atomic Bombs: Estimates of the Genetic Doubling Dose of Radiation for Humans," *Am. J. Hum. Genet.*, 46:1053 (1990).

196. Satoh, C., K. Hiyama, N. Takahashi, M. Kodaira, and J.V. Neel, "Approaches to DNA Methods for the Detection of Heritable Mutations in Humans," in *Mutation and the Environment*, (New York: Wiley-Liss, Inc., 1990), p. 197.

197. Jaffe, D., and G.T. Bowden, "Ionizing Radiation as an Initiator: Effects of Proliferation and Promotion Time on Tumor Incidence in Mice," *Cancer Res.*, 47:6692 (1987).

198. Kaufman, W.K., S.A. Mackenzie, and D.G. Kaufman, "Factors Influencing the Initiation by Gamma Rays of Hepatocarcinogenesis in the Rat," *Teratogenesis Carcinog. Mutagen*, 7:551 (1987).

199. Ariel, M., J. McCarrey, and H. Cedar, "Methylation Pattern of Testis Specific Genes," *Proc. Natl. Acad. Sci. USA*, 88:2317 (1991).

200. Reik, W., A. Collick, M.L. Norris, S.C. Barton, and M.A. Surani, "Genomic Imprinting Determines Methylation of Parental Alleles in Transgenic Mice," *Nature*, 328:248 (1987).

201. Bedford, M.T., and P.D. van Helden, "Hypomethylation of DNA in Pathological Conditions of the Human Prostate," *Cancer Res.*, 47:5274 (1987).

202. Takahashi, T., H. Watanabe, K. Dohi, and A. Ito, "252-Cf Relative Biological

Effectiveness and Inheritable Effect of Fission Neutrons in Mouse Liver Tumorigenesis," *Cancer Res.*, 52:1948 (1992).

203. Nomura, T., "Parental Exposure to X Rays and Chemicals Induces Heritable Tumors and Anomalies in Mice," *Nature*, 296:575 (1982).

204. Kadhim, M.A., D.A. Macdonald, D.T. Goodhead, S.A. Lorimore, S.J. Marsden, and E.G. Wright, "Transmission of Chromosomal Instability After Plutonium Alpha-Particle Irradiation," *Nature*, 355:738 (1992).

205. Chang, W.S.P., and J.B. Little, "Persistently Elevated Frequency of Spontaneous Mutations in Progeny of CHO Clones Surviving X-Irradiation: Association with Delayed Reproductive Death Phenotype," *Mutat. Res.*, 270:191 (1992).

206. Woloschak, G.E., C.M. Chang-Liu, and P. Shearin-Jones, "Regulation of Protein Kinase C by Ionizing Radiation," *Cancer Res.*, 50:3963 (1990).

207. Weichselbaum, R.R., D.E. Hallahan, V. Sukhatme, A. Dritschilo, M.L. Sherman, and D.W. Kufe, "Biological Consequences of Gene Regulation After Ionizing Radiation Exposure," *J. Natl. Cancer Inst.*, 83:480 (1991).

208. Uckun, F.M., L. Tuel-Ahlgren, C.W. Song, K. Waddick, D.E. Myers, J. Kirihara, J.A. Ledbetter, and G.L. Schieven, "Ionizing Radiation Stimulates Unidentified Tyrosine-Specific Protein Kinases in Human B-Lymphocyte Precursors, Triggering Apoptosis and Clonogenic Cell Death," *Proc. Natl. Acad. Sci. USA*, 89:9005 (1992).

209. Nishizuka, Y., "Studies and Perspectives of Protein Kinase C," *Science*, 233:305 (1986).

210. Oh, S.Y., C.G. Grupen, and A.W. Murray, "Phorbol Ester Induces Phosphorylation and Down-Regulation of Connexin 43 in WB Cells," *Biochim. Biophys. Acta*, 1094:243 (1991).

211. Green, A.R., "Recessive Mechanisms of Malignancy," *Br. J. Cancer*, 58:115 (1988).

212. Sun, C., J.L. Redpath, M. Colman, and E. Stanbridge, "Repair of Potentially Lethal Radiation Damage in Confluent and Non-Confluent Cultures of Human Hybrid Cells," *Int. J. Radiat. Biol.*, 49:395 (1986).

213. Dertinger, H., and D. Hulser, "Increased Radioresistance of Cells in Cultured Multicell Spheroids. I. Dependence on Cellular Interaction," *Radiat. Environ. Biophys.*, 19:101 (1981).

214. Dertinger, H., G. Hinz, and K.H. Jakobs, "Intercellular Communication, Three Dimensional Cell Contacts and Radiosensitivity," *Biophys. Struct. Mech.*, 9:89 (1982).

215. Sasaki, T., M. Yamamoto, T. Yamaguchi, and S. Sugiyama, "Development of Multicellular Spheroids of HeLa Cells Cocultured with Fibroblasts and Their Response to X-Irradiation," *Cancer Res.*, 44:345 (1984).

216. Madhoo, J., and G. Blekkenhorst, "Effects of Inhibition of Intercellular Communication on Repair of Damage Induced by Ionizing Radiation," *Brit. J. Radiol.*, 62:281 (1989).

217. West, C.M.L., R.R. Sandhu, and I.J. Stratford, "The Radiation Response of V79 and Human Tumor Multicellular Spheroids-Cell Survival and Growth Delay Studies," *Br. J. Cancer*, 50:143 (1984).

218. Kwok, T.T., and R.M. Sutherland, "The Influence of Cell-Cell Contact on Radiosensitivity of Human Squamous Carcinoma Cells," *Radiat. Res.*, 126:52 (1991).

219. Umezawa, A., K. Harigaya, H. Abe, and Y. Watanabe, "Gap-Junctional Communication of Bone Marrow Stromal Cells Is Resistant to Irradiation In Vitro," *Exp Hematol*, 18:1002 (1990).
220. Bradley, W.E.C., and F. Laviolette, "Low Persistence of the Induced Mutant Phenotype in Chinese Hamster Cells," *Mutat. Res.*, 210:303 (1989).
221. Brouty-Boye, D., I. Gresser, and M.T. Bandu, "Stability of the Phenotypic Reversion of X-Ray Transformed C3H/10T1/2 Cells Depends on Cellular Proliferation After Subcultivation at Low Density," *Carcinogenesis*, 3:1057 (1982).
222. Landolph, J.R., and P.A. Jones, "Mutagenicity of 5-Azacytidine and Related Nucleosides in C3H/10T1/2 Clone 8 and V79 Cells," *Cancer Res.*, 42:817 (1982).
223. Brown, J.D., M.J. Wilson, and L.A. Poirier, " Neoplastic Conversion of Rat Liver Epithelial Cells in Culture by Ethionine and 5-Adenosylethionine," *Carcinogenesis*, 4:173 (1983).
224. Barrett, J.C., and D.G. Thomassen, "Use of Quantitative Cell Transformation Assays in Risk Estimation," in *Methods for Estimating Risk of Chemical Injury: Human and Non-Human Biota and Ecosystems*, B.V. Vouk, G.C. Butler, D.G. Hoel, and D.B. Peakall, Eds., (Chichester, England: John Wiley & Sons, 1985), p. 201.
225. Miller, R.C., and E.J. Hall, "X-Ray Dose Fractionation and Oncogenic Transformation in Cultured Mouse Embryo Cells," *Nature*, 277:58 (1978).
226. Watanabe, M., and K. Suzuki, "Expression Dynamics of Transforming Phenotypes in X-Irradiated Syrian Golden Hamster Embryo Cells," *Mutat. Res.*, 249:71 (1991).
227. Little, J.B., "Influence of Non-Carcinogenic Secondary Factors on Radiation Carcinogenesis," *Radiat Res.*, 87:240 (1981).
228. Darmon, M., J.F. Nicolas, and D. Lamblin, "5-Azacytidine Is Able to Induce the Conversion of Teratocarcinoma-Derived Mesenchymal Cells into Epithelial Cells," *EMBO J*, 3:961 (1984).
229. Trainer, D.L., T. Kline, F. Mallon, R. Greig, and G. Poste, "Effect of 5-Azacytidine on DNA Methylation and the Malignant Properties of B16 Melanoma Cells," *Cancer Res.*, 45:6124 (1985).
230. Jones, P.A., "DNA Methylation and Cancer," *Cancer Res.*, 46:461 (1986).
231. Riggs, A., and P.A. Jones, "5-Methylcytosine, Gene Regulation and Cancer," *Adv. Cancer Res.*, 40:1 (1983).
232. Holliday, R., "DNA Methylation and Epigenetic Defects in Carcinogenesis," *Mutat. Res.*, 181:215 (1987).
233. Tauchi, T., J.H. Ohyashiki, K. Ohyashiki, M. Saito, S. Nakazawa, N. Kimura, and K. Toyama, "Methylation Status of T-Cell Receptor Beta-Chain Gene in B Precursor Acute Lymphoblastic Leukemia: Correlation with Hypomethylation and Gene Rearrangement," *Cancer Res.*, 51:2917 (1991).
234. Hoffman, R.M., "Altered Methionine Metabolism, DNA Methylation and Oncogene Expression in Carcinogenesis," *Biochim. Biophys. Acta*, 738:49 (1984).
235. Clifton, K.H., M.A. Tanner, and M.N. Gould, "Assessment of Radiogenic Cancer Initiation Frequency Per Clonogenic Rat Mammary Cell in Vivo," *Cancer Res.*, 46:2390 (1986).
236. Schwartz, J.L., C.R. Ashman, R.W. Atcher, B.A. Sedita, J.D. Shadley, J.

Tang, J.L. Whitlock, and J. Rotmensch, "Differential Locus Sensitivity to Mutation Induction By Ionizing Radiations of Different LETs in Chinese Hamster Ovary K1 Cells," *Carcinogenesis*, 12:1721 (1991).

237. Nagasawa, H., and J.B. Little, "Induction of Sister Chromatid Exchanges By Extremely Low Doses of X-Particles," *Cancer Res.*, 32:6394 (1992).

238. Cortes, F., W.F. Morgan, and S. Wolff, "Effect of Exogenous Thymidine on Sister Chromatid Exchange Frequency in Chinese Hamster Ovary Cells with Bromodeoxyuridine- and Chlorodeoxyuridine-Substituted Chromosomes," *Mutat. Res.*, 192:277 (1987).

239. Trosko, J.E., C.C. Chang, S.T. Warren, P.K. Liu, M.H. Wade, and G. Tsushimoto, "The Use of Mammalian Cell Mutants to Study the Mechanisms of Carcinogenesis," in *Mutation, Promotion and Transformation in Vitro*, N. Inni, T. Kuroki, M. Yamada, and C. Heidelberger, Eds., (Tokyo: Japan Scientific Soc. Press, 1981), p. 159.

240. Waldren, C., L. Correll, MA. Sognier, and T.T. Puck, "Measurement of Low Levels of X-Ray Mutagenesis in Relation to Human Disease," *Proc. Natl. Acad. Sci. USA*, 83:4839 (1986).

241. Ron, E., B. Modan, D. Preston, E. Alfandary, M. Stovall, and J.D. Boice, Jr., "Radiation-Induced Skin Cancers of the Head and Neck," *Radiat. Res.*, 125:318 (1991).

242. Richardson, R.B., "Past and Revised Risk Estimates for Cancer Induced By Irradiation and Their Influence on Dose Limits," *Brit. J. Radiol.*, 63:235 (1990).

243. Little, M.P., M.M. Hawkins, M.W. Charles, and N.G. Hildreth, "Fitting the Armitage Doll Model to Radiation-Exposed Cohorts and Implications for Population Cancer Risks," *Radiat. Res.*, 132:207 (1992).

244. Yoshimoto, Y., "Cancer Risk Among Children of Atomic Bomb Survivors," *J. A. M. A.*, 264:596 (1990).

245. Thompson, D.E., K. Mabuchi, E. Ron, M. Soda, M. Tokunaga, S. Ochikubo, S. Sugimoto, T. Ikeda, M. Terasaki, S. Izumi, and D.L. Preston, "Cancer Incidence in Atomic Bomb Survivors. Part II: Solid Tumors, 1958–1987," *Radiat. Res.* 137S17-S67 (1994).

246. Land, C.E., M. Tokunaga, and S. Tokuoka, "Studies of Breast Cancer Risks Among Atomic Bomb Survivors," in *Brit. Inst. Radiol. Report 22*, G.B. Gerber, D.M. Taylor, E. Cardis, and J.W. Thiessan, Eds., London, 1991, p. 49.

247. Brash, D.E., J.A. Rudilph, J.A. Simon, A. Lin, G.J. McKenna, H.P. Baden, A.J. Halperin, and J. Portan, "A Role for Sunlight in Skin Cancer: UV-Induced p53 Mutations in Squamous Cell Carcinoma," *Proc. Natl. Acad. Sci. USA*, 88:10124 (1991).

248. Shore, R.E., "Overview of Radiation-Induced Skin Cancer in Humans," *Int. J. Radiat. Biol.*, 57:809 (1990).

CHAPTER 15

Stress Proteins and Radiation

Joan Smith-Sonneborn, Zoology & Physiology Department,
University of Wyoming, Laramie, Wyoming

INTRODUCTION

Hormesis is the phenomenon of induction of beneficial effects by low doses of otherwise harmful physical or chemical agents observed with diverse physical and chemical agents.[1] The present report presents evidence of beneficial biological responses to low doses of physical and chemical agents and a theory to test whether a fundamental principle may be operative to explain the results.

To explore the hypothesis that the hormetic response may operate through the stress response, this report includes: (1) identification of agents known to induce both the stress response and hormetic phenomena; (2) a biological response to radiation and electromagnetic fields; (3) a description of the unique and common pathways in the stress response to three stressors: heat, DNA damaging agents, and teratogens; (4) a proposed explanation for the paradoxical beneficial response to low doses of an otherwise harmful agent via a stress-response pathway; and (5) the molecular evidence for the induction of the stress response to low levels of radiation.

STRESS AND HORMETIC AGENTS

A physiological basis for a common pathway[2,3] for the beneficial low dose response is the induction of a "heat shock-like" response.[4,5] The heat shock response is a model for a more general phenomenon, called "the stress response." The stress response is characterized by increased synthesis of a family of stressor-specific proteins with concomitant reduction of synthesis of most other proteins. The stress response has been characterized using heat, radiation, heavy metals, and oxidizing agents as the stressors.[6-8]

Chemical and physical hormetic agents include heavy metals, polychlorinated biphenyls, insecticides, alcohol, oxygen poisoning, cyanide, antibiotics, ionizing radiation,[9] cosmic or gamma radiation,[1,9] electromagnetic radiations,[10-12] and ultraviolet plus photoreactivation.[13] Examples of beneficial biological responses include increased life span, cell division rate,[10-12] accelerated maturation time, and acclimation.[8]

Agents identified as both hormetic agents and inducers of the stress response include cadmium, mercury, copper, zinc, ethanol, oxygen, chloramphenicol, X-rays, ultraviolet light plus photoreactivation, and heat.[1,4,5,8] There is an impressive overlap between the hormetic and stress response inducers, although previous experiments did not correlate the stress response and the onset of a beneficial biological effect. However, a biomarker of the stress response is induced resistance to the stressor. Recent studies do confirm the induction of a stress response, and focus on the molecular bases of the adaptation to the damage.[9]

RADIATION EFFECTS

Ultraviolet Irradiation

The life span of the ciliated protozoan, *Paramecium*, was extended by exposure to low doses of ultraviolet irradiation followed by photoreactivation. High doses of ultraviolet were lethal. The dose required to significantly damage cells decreased with age as did the dose required to induce the beneficial response in the paired damage and repair cycle. Prior exposure to irradiation-induced resistance and resulted in a more youthful tolerance to irradiation in the older cells.[13]

Another ciliate, *Tetrahymena*, shows an incremental increase in DNA polymerase with increase in radiation. Both mitochondrial and nuclear DNA polymerases responded to the damage. A repressor of DNA polymerase was thought destroyed by the radiation.[14]

Extremely Low Levels of Electromagnetic Radiations

The effect of extremely low electromagnetic radiations on the ciliate *Paramecium* has been investigated using alteration of swimming patterns, cell division rate, and longevity as the biological markers.

Paramecium has an excitable membrane which shares many properties common to both nerve and muscle cells including an action potential.[15] A calcium influx mediates depolarization which, in paramecia, is coupled with reversal in the ciliary axoneme forward motion, and results in temporary backward swimming. This "avoidance reaction" is a graded response (rather than "all or nothing"), so is based on the strength of the applied stimulus.[16] While the calcium channel governing calcium influx is voltage dependent and hence, a "fast" channel with millisecond kinetics, the sodium channel is a "slow" channel (0.1 second kinetics). The sodium channel prolongs the depolarization response in both mutant and wild type paramecia.[17] Behavioral responses could be categorized into different swim patterns. The threshold levels of sodium concentrations required to alter the swimming pattern in ion baths has been used as a biological marker of agent-induced cell surface changes. Frequency dependent alterations in swimming patterns were found.[11] At 15 hertz, higher concentrations of sodium were required to induce backward swimming compared with controls, while at 25 hertz, lower concentrations of sodium, relative to controls were required to induce the swimming pattern change.[11]

Paramecia have also been used to detect biological effects of electromagnetic fields.[10-12] The extremely low frequency electromagnetic fields were produced using horizontally positioned twin air gap Helmholtz coils driven by a programmable wave generator supplied by Electro-biology Inc., Fairfield, N.J., with current signals designed to couple to electrochemical processes at cell surfaces.[19] Both single pulse and pulse burst wave forms were used with frequencies from 15 to 75 hertz.

Increased *Paramecium* longevity relative to controls was found when the cells were cultivated in specific electromagnetic signals between 15 to 75 hz given at specific age intervals.[10] The site of the biological effects on these ciliates was presumed to be the cell membrane.

That a biological response of the electromagnetic effects exists at the surface is supported by: (1) alteration in fluorescent probe binding which detect changes in membrane and the membrane fluidity in response by these electromagnetic fields,[12] and (2) frequency-dependent induced changes in threshold sodium concentrations are required to reverse the swimming pattern. Previous studies emphasize the importance of frequency on biological effects using calcium efflux changes in brain cells.[20,21] The similarities between the results with paramecia and nerve cells emphasize the use of paramecia as a "swimming neuron".[15]

Thus, low levels of ultraviolet, electron, and electromagnetic irradiation induced changes which can be detected at the cells' surface, cytoplasm, and nucleus. In the quest for a common mechanism to explain the paradoxical responses of low and high levels of agents, the role of stress response has been targeted for investigation.

THE STRESS RESPONSE

The stress proteins are divided into two groups: those referred to as the heat shock proteins first found induced by nonphysiological exposure to heat; and those called the glucose-regulated proteins, which exhibit increased synthesis when cells are deprived of glucose, oxygen, or disruption of calcium homeostasis. Members of the two families exhibit considerable homology.[6]

Heat

The first stress response detected was the appearance of puffs induced in the salivary gland of fruit flies by heat.[22] The heat shock response is universal from bacteria to man.[23] The stress response induced by heat is characterized by transcription of coordinately regulated subset of induced proteins, repression of the transcription, and translation of previously active genes and preexisting messages. The heat shock proteins are members of families of proteins with species-related molecular weight range classes of proteins. In higher organisms, the high molecular weight families are: Hsp110, a normal nucleolar protein found in vertebrates; the Hsp 100 family (Mr 92-102), a phosphoprotein normally present in the plasma membrane; Hsp 89 (Mr 83-95), found in the soluble protein in all animal cells; and the Hsp 70 family (Mr 68-78).[6,24] The multigene hsp70 family has members in the cytoplasm, in the lumen of the endoplasmic reticulum, and in the matrix of the mitochondria which function in protein translocation across membranes.[24-26]

The heat shock protein hsp70 is a major factor in the heat response since (1) mammalian cells in which hsp70 is not made or is inactivated by antibody binding cannot develop thermotolerance[26]; (2) cycloheximide can induce tolerance to higher temperatures, but heat shock proteins are required for full protection[27]; and (3) the thermolability of mouse oocytes is due to the lack of expression and/or inducibility of hsp70.[28]

Smaller hsps (15-28 kd) bind reversibly to the nuclear skeleton during heat shock and form higher order aggregates. A common central domain of the four small Drosophila hsps (22, 23, 26,28) show great similarity to alpha crystallin.[25,29]

The smallest hsps are the ubiquitin family 7-8 kd, which have been implicated as regulator molecules in chromatin,[30-36] DNA repair,[36] meiosis, sporulation, degradation of abnormal proteins,[31] ribosome biogenesis,[32] and facilitation of transposition.[34]

Besides the induction of heat shock proteins, other metabolic changes found in response to nonphysiological heat exposure which impact on chromatin structure include: (1) increased levels of high molecular weight ubiquitin conjugates and decreased ubiquitinated histone in HeLa cells[34,35] (ubiquitinated DNA is associated with active expression);[36] (2) hypermethylation of H2B and decreased methylation of H3;[37] and (3) the ubiquitinated form of histones in yeast when grown under mildly stressful but not lethal temperature.[34]

Topological changes in chromatin are typical of the heat shock response[37] and are assumed to participate in the changes in heat-induced gene expression and repression.

A presumed physiological consequence of heat-induced alteration of chromatin is heat-induced radiosensitivity of cancer cells.[38,39] Heat induces a dramatic increase of nonhistone protein content, resulting in a reduced affinity to repair enzymes.[38]

Heat also causes conformational changes of membrane lipids and proteins,[40,41] excessive fluidization of the plasma membrane, and leakage of required low molecular weight components.[39] Low doses of local anesthetics, procaine and lidocaine, are known to decrease membrane viscosity and increase neoplastic killing.[41] The membrane defects may cause release of polyamines and disturb DNA replication.[38,39]

DNA-Damaging Agents

In prokaryotes, response to a given stressor, unlinked and individually controlled genes can be coordinately controlled by common regulator genes called regulons.[6] The damage response in bacteria to ultraviolet irradiation is the "SOS" response[42,43] to reactive oxygen species, the Oxy R response,[44] and specialized responses to other environmental stresses like cold, heat, nutrient limitation, salinity, and osmolarity are well characterized.[45] Different stressors are related in the sense that they share member genes or protein products that interact. For example, in *Escherichia coli*, both heat and ethanol initiate the same response; i.e., solely a heat shock response. On the other hand, both hydrogen peroxide and 6 amino-7-chloro-5,8-dioxoquinoline (ACDQ) stimulate an oxidation stress response and a secondary SOS response; nalidixic acid and puromycin, an SOS and heat shock response; isoleucine restriction, a poor heat shock response; and cadmium chloride strongly induces all three stress responses.[7] The regulon-typical response to ACDQ, cadmium chloride, and hydrogen peroxide was a minor response; these agents stimulated the synthesis of another 35 proteins by 5- to 50-fold. Another accumulated product after exposure to certain stressors is adenylated nucleotides, the so-called "alarmones."[7]

Thus, general and specific cellular responses are triggered by different stressors. Ultraviolet or carcinogen-related DNA damage-induced expression of the stress response does not appear to conform to the prokaryote SOS model.[46] DNA damaging agents induce a spectrum of molecular responses, including the production of proteases, DNA repair agents, oncogenes, and chromatin changes. DNA damage is associated with the induction of specific identified (and unidentified) proteins, including a plasminogen activator,[47] increased synthesis of poly(ADP-ribose),[48] DNA ligase,[49] an SOS-like response in mammalian cells,[50-53] alteration of histone methylation patterns,[54] and dependence on the presence of ubiquitin-histone conjugants.[30] In contrast with heat shock, DNA-damaging agents inhibit rather than increase DNA methylation.[54] The ubiquitin-conjugating enzyme is essential for DNA repair, since loss of the ubiquitin-conjugating enzyme, E3, results in slow growth; sensitivity to UV, X-rays, and chemical mutagens; retrotransposition; and inability to sporulate.[30] A suggested role for the ubiquitin-conjugating enzyme is to mediate changes in chromatin by patched degradation of chromosomal proteins to allow access for repair.[30]

Degradation of Abnormal Proteins Produced by Stressors

When cells are exposed to heat and other toxic agents, abnormal proteins accumulate. The abnormal proteins signal expression of heat shock proteins which can directly interact with the protein for "protein repair" by catalyzing a ATP-driven refolding.[55] The unrepaired proteins are eliminated by a second major pathway of the response, an ATP-driven elimination of abnormal proteins mediated by the ubiquitin system. However, imbalances in the protein degradation system, perhaps induced by an overload of abnormal proteins, can result in premature degradation of necessary regulatory molecules.[56] With respect to hormesis, the beneficial stress response may be protein repair and the elimination of abnormal proteins. The detrimental response may be the inappropriate destruction of short-lived essential regulator molecules when a threshold level of abnormal proteins is produced by toxic agents or radiation damage, aging, or age-related diseases. In addition to imbalance in the degradation pathway at higher doses when abnormal proteins accumulate, changes in the fundamental structure of the essential ubiquitin-conjugating enzymes[30] is dosage-dependent, at least with respect to heat.[30] The ubiquitin-conjugating enzymes, essential for survival to the stressing agent, have introns. Since splicing of introns is blocked at higher temperatures,[57,58] the introns could serve to restrict a gene protective mediated proteolysis to moderate, not severe stress.[31]

MOLECULAR EVIDENCE FOR INDUCTION OF THE STRESS RESPONSE

DNA Damage and Stress Proteins

The role of heat shock genes in the DNA damage response is not known, but heat shock genes do appear in the DNA damage response.[59] The members of the heat shock protein family identified in response to DNA damaging agents are hsp70 and small hsps induced by UV and teratogenic agents.[59,60] The small hsp, ubiquitin, was induced after treatment with mutagens and teratogens.[61-65]

Using ionizing irradiation of rat embryos in utero, enhanced expression of hsp70 and c-myc was increased 4 or 5 days after treatment, and c-fos increased only after the embryos were incubated in vitro.[65] Coordinate expression of hsp70 and c myc has been detected during heat shock.[66]

In human cells in culture, certain X-ray inducible genes and their corresponding proteins appearing after priming doses are candidates for induction of radioresistance, especially a 269 KDa protein.[67] In mice, chronic

X-ray irradiation was found to increase the expression of hsp70 at a dose rate of 3 cGy/day over a 30–60 day interval, especially in spleen cells of mice.[68] Increased hsp70 mRNA and protein were also found with specific low dose delivery rates in spleen cell, and chronic low doses of ionization could influence the progression of autoimmune disease.[69,70]

In yeast cells, gamma radiation-induced resistance was dependent on the total radiation dose rather than the dose rate.[71] A hsp104 deficient mutant could still become radio-resistant, indicating that this hsp104 was not required for heat shock induced radio-resistance.

Chemical teratogens showed enhanced induction of small heat-shock proteins in embryos when cultivated in vitro,[65] in flies,[72] and the small hsp, ubiquitin, in mammalian cells.[73,74]

Since the ubiquitin-conjugated Rad6 DNA repair enzyme is a ubiquitin-conjugating enzyme[30] essential for normal growth, sporulation, and repair, ubiquitin may be a key regulatory molecule in the stress response. Changes in metabolism of ubiquitin as well as increased synthesis of unique forms of ubiquitin gene family members may shed light on controlling elements.

STRESS AND HORMESIS: SUMMARY

The hypothesis that the stress protein response is the common pathway for hormetic agents is supported by the finding that: (1) the same agents identified as hormetic also induce the stress response; (2) some hormetic agents with molecular responses common to the heat shock response can induce thermotolerance, while others with known differences in induction of methylation patterns induce sensitivity; (3) the stress response includes preferential synthesis of products which repair both protein and DNA, which could stimulate growth and longevity; (4) the alterations in the chromatin structure could facilitate derepression of growth-promoting products or provide access to DNA for repair; and (5) low chronic doses induce the known stress proteins and other adaptive agents. There is a model for a biphasic response using heat as the stressor. At moderate doses, the protective molecular reactions progress, but at higher temperatures, intron splicing is inhibited, and therefore, production of the needed protective response. Other, as yet unknown important differences in molecular responses at low and high doses may be uncovered in the future.

In summary, the stress response could provide an explanation for a beneficial response to an otherwise harmful agent. The potential for a theoretical biological beneficial response stems from the induction of cellular repair processes. The protective responses include: (1) expression of "protein repair" proteins, like the heat shock proteins, which can monitor proper folding of denatured proteins; (2) stimulation of elimination of abnormal proteins which cannot be repaired; (3) induction of increased DNA repair and DNA replication molecules; (4) alteration of chromatin structure to facilitate repair of regions previously refractory to repair, or to alter gene expression to accelerate growth and maturation; and (5) induction of cross-resistance to other environmental toxins, thereby increasing tolerance to the same or apparently unrelated environmental toxins which are life-shortening agents.

Why the beneficial response is affected only at low doses cannot yet be explained, but the inability to remove introns from gene transcripts required for survival, at moderate but not high temperatures, and changes in histone-ubiquitin conjugates may provide clues to explain cytotoxic and genotoxic responses after a threshold limit for a beneficial response.

Since different stressors have specific responses, not all stressors are expected to be beneficial or beneficial with respect to the same parameter. The hormetic response may not be an "overcorrection" response to the damaging agent, but rather a benefit derived from the "stress response"; i.e., repair or removal of accumulated age- or environmental-induced cellular damage in proteins, genes, and cell membranes, chromatin changes to accelerate seed maturation, or cross-resistance to certain other environmental toxins.

REFERENCES

1. *Health Physics*, 52:517–680 (1987).
2. Luckey, T. D., "Ionizing Radiation Promotes Protozoan Reproduction," *Radiat. Res.* 108:215–221 (1986).
3. Stebbing, A. R. D., "Growth Hormesis: A Byproduct of Control," *Health Physics*, 52:543–548 (1987).
4. Smith-Sonneborn, J., "The Role of the Stress Protein Response in Hormesis," in *Biological Effects of Low Level Exposure to Chemicals and Radiation*, E. J. Calabrese, Ed., (Chelsea, MI: Lewis Publishers, 1992), pp. 41–52.
5. Smith-Sonneborn, J., "The Role of the Stress Protein Response in Hormesis," in *Low Dose Irradiation and Biological Defense Mechanisms*, T. Sugahara, L.A. Sagan, and T. Aoyama, Eds. (New York: Excerpta Medica, 1992), pp. 399–404.
6. Welch, W. J., L.A. Mizzen, and A.P. Arrigo, "Structure and Function of Mammalian Stress Proteins," in *Stress-Induced Proteins*, M.L. Pardue, J.R. Feramisco, and S. Lindquist, Eds., (New York: Alan R. Liss, 1989), p. 187.
7. VanBogelen, R. A., P.M. Kelley, and F.C. Neidhardt, "Differential Induction of Heat Shock, SOS and Oxidation Stress Regulons and Accumulation of Nucleotides in *Escherichia coli*," *J. Bacteriol.* 169:26–32 (1987).
8. Calabrese, E. J., M.E. McCarthy, and E. Kenyon, "The Occurrence of Chemically Induced Hormesis," *Health Physics*, 52:531–542 (1987).
9. *Low Dose Irradiation and Biological Defense Mechanisms*, T. Sugahara, L.A. Sagan, and T. Aoyama, Eds., (Amsterdam: Elsevier Science Publishers, 1992).
10. Smith-Sonneborn J., "Programmed Increased Longevity Induced by Weak Pulsating Current in *Paramecium*," *Bioelectrochem. Bioenerg.*, 11:373–382 (1983).
11. Darnell, C., "Effects of Extremely Low Electromagnetic Radiation on *Parame-*

cium Lifespan and Ion Conductance," M.S. thesis, University of Wyoming, Laramie, WY, 1988.

12. Dihel, L., and J. Smith-Sonneborn, "Effects of Low Frequency Electromagnetic Fields on Cell Division and the Plasma Membrane," *Bioelectromagnetics*, 6:61–71 (1985).

13. Smith-Sonneborn, J., "DNA Repair and Longevity Assurance in *Paramecium tetraurelia*," *Science*, 203:1115–1117 (1979).

14. Keiding, J., and O. Wesstergaard, "Induction of DNA Polymerase Activity in Irradiated *Tetrahymena* Cells," *Exp. Cell Res.*, 64:317–322 (1971).

15. Kung, C., "Calcium Channels of *Paramecium*: A Multidisciplinary Study," in *Current Topics in Membranes and Transpor.*, 23:45–66 (1985).

16. Eckert, R., Y. Naitoh, and H. Machemer, "Calcium in the Bioelectric and Motor Functions of *Paramecium*," *Calcium in Biological Systems, Cold Spring Harbor Symp. Soc. Exp. Biol.* 30:233–255 (1976).

17. Adoutte A., K.Y. Ling, M. Forte, R. Ramanathan, D. Nelson, C. Kung, "Ionic Channels of *Paramecium*: From Genetics and Electrophysiology to Biochemistry," *J. Physiol., (Paris)*, 77:1145–1159 (1981).

18. Chang, S.Y, and C. Kung, "Selection and Analysis of a Mutant *Paramecium tetraurelia* Lacking Behavioral Response to Tetraethylammonium," *Genetic Res.*, 27:97–107 (1976).

19. Pilla, A.A., J.J. Kaufman, and G. Rein, "Electrochemical Mechanisms in the Electromagnetic Modulation of Tissue Growth and Repair," *J. Elchem. Soc.*, 133(3):c124 (1986).

20. Bawin, S.M, W.R. Adey, and I.M. Sabbott, "Ionic Factors in Release of Calcium from Chicken Cerebral Tissue by Electromagnetic Fields," *Proc. Natl. Acad. Sci. USA*, 75:6314–6318 (1978).

21. Blackman, C.F., S.G. Berane, D.E. House, and W.T. Joines, "Effects of ELF (1–120 Hz) and Modulated (50 Hz) RF Fields on the Efflux of Calcium Ions from Brain Tissue in Vitro," *Bioelectromagnetics*, 6:1–11 (1985).

22. Ritossa, F. M., "A New Puffing Pattern Induced by Heat Shock and DNP in *Drosophila*," *Experientia*, 18:571–573 (1961).

23. Nover, L., and K.D. Scharf, *Heat Shock Response*, (New York: CRC Press, 1991), pp. 41–127.

24. Lindquist, S., and E.A. Craig, "The Heat Shock Proteins," *Ann. Rev. Genet.*, 22:631–677 (1988).

25. Craig, E., P.J. Kang, and W.A. Boorstein, "A Review of the Role of 70 kDa Heat Shock Proteins in Protein Translocation Across Membranes," *Antonie von Leeuwenhoek*, 58:137–146 (1990).

26. Riabowol, K.T., L.A. Mizzen, and W.J. Welch, "Heat Shock is Lethal to Fibroblasts Microinjected with Antibodies Against HSP 70," *Science*, 243:433–436 (1988).

27. Hallberg, R.I., K.W. Kraus, and E.M. Hallberg, "Induction of Acquired Thermotolerance in *Tetrahymena thermophila* Can Be Achieved Without the Prior Synthesis of Heat Shock Proteins," *Mol. Cell. Biol.*, 5:2061–2070 (1985).

28. Hendrey, J., and I. Kola, "Thermolability of Mouse Oocytes is Due to the Lack of Expression and/or Inducibility of Hsp70," *Mol. Reproduc. Devel.*, 28:1–8 (1991).

29. Tuite, M. F., N.J. Bentley, and P. Bossier, "The Structure and Function of

Small Heat Shock Proteins: Analysis of the *Saccharomyces cerevisiae* Hsp26 Protein," *Antonie van Leeuwenhoek*, 58:147–154 (1990).

30. Jentch, S., W. Seufert, and T. Sommer, "Ubiquitin-Conjugating Enzymes: Novel Regulators of Eukaryotic Cells," *TIBS*, 15:195–198 (1990).

31. Dice, J.F., and S.A. Goff, "Error Catastrophe and Aging: Future Directions of Research," in *Modern Biological Theories of Aging*, H.R. Warner, R.N. Butler, R.L. Sprott, and E.L. Schneider, Eds., (New York: Raven Press, 1987), pp. 155–168.

32. Finley, D., B. Bartel, and A. Varshavsky, "The Tails of Ubiquitin Precursors are Ribosomal Proteins Whose Fusion to Ubiquitin Facilitates Ribosome Biogenesis," *Nature*, 338:394–401 (1989).

33. Picologou, S., N. Brown, and S.W. Liebman, "Mutations in RAD6, A Yeast Gene Encoding a Ubiquitin-Conjugating Enzyme, Stimulate Retrotransposition," *Mol. Cell. Biol.*, 10(3):1017–1022 (1990).

34. Pratt, G., Q. Deveraux, and M. Rechsteiner, "Ubiquitin Metabolism in Stresses Cells," in *Stress-induced Proteins*, M.L. Pardue, J.R. Feramisco, and S. Lindquist, Eds., *UCLA Symposium on Molecular and Cellular Biology*, 96:149–162 (1989).

35. Bonner, W. M., "Metabolism of Ubiquitinated H2A," in *The Ubiquitin System*, M. Schlesinger and A. Hershko, Eds., (Cold Springs, NY: Cold Springs Harbor Laboratory, 1988), pp. 155–158.

36. Davie, J. R., S.E. Nickel, and J.A. Ridsdale, "Ubiquitinated Histone H2B is Preferentially Located in Transcriptionally Active Chromatin," in *The Ubiquitin System*, M. Schlesinger and A. Hershko, Eds., (Cold Springs Harbor, NY: Cold Springs Harbor Laboratory, 1988), pp. 159–163.

37. Higgins, C.F., "DNA Supercoiling, Chromatin Structure and the Regulation of Gene Expression," *Antonie van Leeuwenhoek*, 58:51–55 (1990).

38. Carper, S.W., P.M. Harari, and D.J.M. Fuller, "Biochemical and Cellular Response to Hyperthermia in Cancer Therapy," in *Stress-Induced Proteins*, M.L. Pardue, J.R. Feramisco, and S. Lindquist, Eds., *UCLA Symposium on Mol. Cell. Biol.*, 96:247–256 (1989).

39. Hahn, G.M., M.K.I. Adwankar, and V.S. Basrur, "Survival of Cells Exposed to Anticancer Drugs After Stress," in *Stress-Induced Proteins*, M.L. Pardue, J.R. Feramisco, and S. Lindquist, Eds., *UCLA Symposium on Molecular and Cellular Biology*, (New York: Alan Liss, 1989), pp. 223–234.

40. Lepock, J.R., K.H. Cheng, and J. Kruuv, "Thermotropic Lipid and Protein Transitions in Chinese Hamster Lung Cell Membranes: Relationship to Hyperthermic Cell Killing," *Can. J. Biochem. Cell Biol.*, 61:421–427 (1983).

41. Yatvin, M. B., "The Influence of Membrane Lipid Composition and Procaine on Hyperthermic Death of Cells," *Int. J. Radiat. Biol.*, 32:513–521 (1977).

42. Witkin, E.M., "Ultraviolet Mutagenesis an Inducible DNA Repair in *Escherichia coli*," *Bacteriol. Rev.*, 40:869–907 (1976).

43. Walker, G. C., "Mutagenesis and Inducible Responses to Deoxyribonucleic Acid Damage in *Escherichia coli*," *Microbiol. Rev.*, 48:60–93 (1984).

44. Storz, G., L.A. Tartaglia, and B.N. Ames, "The OxyR Regulon," *Antonie van Leeuwenhoek*, 58:157–161 (1990).

45. Bhagwat, A.A., and S.K. Apte, "Comparative Analysis of Proteins Induced by Heat Shock, Salinity and Osmotic Stress in the Nitrogen-Fixing Cyanobacterium, *Anabaena* sp. Strain L31," *J. Bacteriology*, 171:5187–5189 (1989).

46. Elespuru, R.K., "Inducible Responses to DNA Damage in Bacteria and Mammalian Cells," *Environ. Mol. Mutagen.*, 10:97–116 (1987).
47. Miskin, R., and E. Reich, "Plasminogen Activator: Induction of Synthesis by DNA Damage," *Cell.*, 19:217–224 (1980).
48. Ueda, K., and O. Hayaishi, "ADP-Ribosylation," *Ann. Rev. Biochem.*, 54:73–100 (1985).
49. Mezzina, M., and S. Nocentini, "DNA Ligase Activity in UV-Irradiated Monkey Kidney Cells," *Nucleic Acids Res.*, 5:4317–4334 (1978).
50. Sarasin, A., "SOS Response in Mammalian Cells," *Cancer Invest.* 3(2):163–174 (1985).
51. Herrlich, P., P. Angel, and H.J. Rahmsdorf, "The Mammalian Genetic Stress Response," *Adv. Enzyme Regul.*, 25:485–504 (1986).
52. Herrlich, P., U. Mallick, and H. Ponta, "Genetic Changes in Mammalian Cells Reminiscent of an SOS Response," *Hum. Genetics*, 67:360–368 (1984).
53. Lieberman, M.W., L.R. Beach, and R.D. Palmiter, "Ultraviolet Radiation Induced Metallothionein-Igene Activation Is Associated with Extensive DNA Demethylation," *Cell.*, 5:207–214 (1983).
54. Wilson, V.L., and P.A. Jones, "Inhibition of DNA Methylation of Chemical Carcinogens in Vitro," *Cell.*, 32:239–246 (1983).
55. Rothman, J.E., "Polypeptide Chain Binding Proteins: Catalysts of Protein Folding and Related Processes in Cells," *Cell.*, 59:591–601 (1989).
56. Dice, J.F., "Lysosomal Pathways of Protein Degradation," in *The Ubiquitin System*, M. Schlesinger and A. Hershko, Eds., (Cold Springs, NY: Cold Spring Harbor Laboratory, 1988), pp. 141–146.
57. Yost, H.J., and S. Linquist, "RNA Splicing Is Interrupted by Heat Shock and Is Rescued by Heat Shock Protein Synthesis," *Cell.*, 45:185–193 (1986).
58. Yost, H.J., and S. Lindquist, "Translation of Unspliced Transcripts After Heat Shock," *Science*, 242:1544–1548 (1988).
59. Higo, H., J.Y. Lee, Y. Satow, and J. Higo, "Elevated Expression of Protoncogenes Accompany Enhanced Induction of Heat-Shock Genes After Exposure of Rat Embryos in Utero to Ionizing Irradiation," *Teratogenesis, Carcinogenesis, and Mutagenesis*, 9:191–198 (1989).
60. Buzin, C.H., and N. Bournias-Vardiabasis, "Teratogens Produce a Subset of Small Heat Shock Proteins in *Drosophila* Primary Embryonic Cell Cultures," *Proc. Natl. Acad. Sci.*, 81:4075–4079 (1984).
61. Friedberg, E.C., "Deoxyribonucleic Acid Repair in the Yeast *Saccharomyces cerevisiae*," *Microbiol. Rev.*, 52:70–102 (1988).
62. N. Bournias-Vardiabasis, R.L. Teplitz, G.F. Chernoff, and R.L. Seecof, "Detection of Teratogens in *Drosophila* Embryonic Cell Culture Test Assay of 100 Chemicals," *Teratology*, 28:109–122 (1983).
63. Tregar, J.M., K.A. Heichman, and K. McEntee, "Expression of the Yeast UB14 Gene Increases in Response to DNA-Damaging Agents and in Meiosis," *Mol. Cell. Biol.*, 8:1132–1136 (1988).
64. McClanahan, T., and K. McEntree, "DNA Damage and Heat Shock Dually Regulate Genes in *Saccharomyces cerevisiae*," *Mol. Cell. Biol.*, 6:90–96 (1986).
65. Higo, H., K. Higo, and J.Y. Lee, "Effects of Exposing Rat Embryos in Utero to Physical or Chemical Teratogens Are Expressed Later As Enhanced Induction of Heat Shock Proteins When Embryonic Hearts Are Cultivated in Vitro," *Teratogenisis, Carcinogenisis and Mutagenesis*, 18:315–328 (1988).

66. Kingston R.E., A.S. Baldwin, Jr., and P.A. Sharp, "Regulation of Heat Shock Protein 70 Gene Expression by c-myc," *Nature*, 12:280–282 (198)4.

67. Meyers, M., R.A. Schea, A.E. Petrowski, H. Seabury, P.W. McLaughlin, I. Lee, S.W. Lee, and D.A. Boothman, "Role of X-Ray-Inducible Genes and Proteins in Adaptive Survival Responses," in *Low Dose Irradiation and Biological Defense Mechanisms*, T. Sugahara, L.A. Sagan, and T. Aoyama, Eds., (Amsterdam: Exerpta Medica, Elsevier Science Publishers, 1992), pp. 263–266.

68. Melkonyan, H., T. Ushakova, and S. Umansky, "Chronic Irradiation Increases the Expression of Hsp70 Genes in Mouse Tissues," in *Low Dose Irradiation and Biological Defense Mechanisms*, T. Sugahara, L.A. Sagan, and T. Aoyama, Eds., (Amsterdam: Exerpta Medica, Elsevier Science Publishers, 1992), pp. 243–236.

69. Nogami, H., J.M. Lubinski, and T. Makinodan, "Stress Proteins Induced by Chronic Low Dose Ionizing Radiation (LDR) in Vivo Is Correlated with Augmentation of Mitogenic Response in Murine Spleen Cells," *J. Cell Biochem.*, 1991, Suppl. 15F 77.

70. Makinodan T., "Cellular and Subcellular Alterations in Immune Cells Induced by Chronic, Intermittent Exposure in Vivo to Very Low Doses of Ionizing Radiation (LDR) and Its Ameliorating Effects on Progression of Autoimmune Disease and Mammary Tumor Growth," in *Low Dose Irradiation and Biological Defense Mechanisms*, T. Sugahara, L.A. Sagan, and T. Aoyama, Eds., (Amsterdam: Exerpta Medica, Elsevier Science Publishers, 1992), pp. 233–237.

71. Boreham, D.R., M.E. Bahen, S. Laffrenier, and R.E.J. Mitchel, "Stress-Induced Radiation Resistance in Yeast," in *Low Dose Irradiation and Biological Defense Mechanisms*, T. Sugahara, L.A. Sagan, and T. Aoyama, Eds., (Amsterdam: Exerpta Medica, Elsevier Science Publishers, 1992), pp. 267–270.

72. Petersen, N.S., "Effects of Heat and Chemical Stress on Development," *Advances in Genetics*, 28:275–296 (1990).

73. Fornace, A., I. Alamo, and C.M. Hollander, "Induction of Heat Shock Protein Transcripts and B2 Transcripts by Various Stresses in Chinese Hamster Cells," *Exp. Cell Res.*, 182:61–74 (1989).

74. Fornace, A., Jr., I. Hollander, and C. Hollander, "Ubiquitin mRNA is a Major Stress-Induced Transcript in Mammalian Cells," *Nucleic Acid Research*, 17:1215–1229 (1989).

CHAPTER 16

Adaptive Responses After Exposure to Low Dose Radiation

Colin K. Hill and **Tony Godfrey**,
Department of Radiation Oncology and Microbiology,
Albert Soiland Cancer Research Laboratory,
University of Southern California School of Medicine, Los Angeles, California

INTRODUCTION

Since the exploding of the first atomic bomb at the end of the second World War there has been an overriding and all-pervading concern about the deleterious effects of radiation. These concerns have led to radiation sources of all types being, perhaps, the most closely regulated of potentially toxic substances in modern society. However, despite this notoriety, radiation of various kinds occupies an important niche in our society, particularly in medicine, where x-rays and nuclear isotopes, for example, have vastly improved diagnostic abilities and thus the quality of our health in general.

The efforts of physicists and chemists have given us methods for detecting truly minute quantities of radiation. The result, however, has not always been a wise use of this knowledge, because some professionals and some lay people call for any detectable speck of radiation to be removed or outlawed from our environment. Perhaps because of the specter of nuclear holocaust and the horrors of radiation sickness, emotions often run higher than common sense would seem to justify. Thus, there is much effort, time, and money spent on determining the levels of radiation in every facet of our environment, how it affects our health, and what its long-term consequences or risk to humans will be.

In among all the scientific reports of deleterious effects of radiation there have been a growing number of scientific, and occasionally not so scientific, reports that radiation can sometimes be "beneficial." Although the use of the word beneficial is perhaps misleading as it suggests radiation can be benign and good, the reports include a number of studies clearly showing alterations in cell, tissue, or organ response to radiation that are not immediately deleterious.

In recent years increased attention has been given to small doses of radiation and to the 'stress' responses they seem to elicit. Molecular techniques are now showing that radiation, among many other outside influences, can cause genes to be activated and new proteins to be produced in cells. From these studies has come the term "adaptive response," connoting that: (1) the cell, tissue, or organism changes in a way to *improve* its health (i.e., chance of survival) in the presence of radiation; (2) the cell, tissue, or organism produces something, presumably proteins, in response to radiation that was not normally present, and that these proteins effect the "beneficial changes."

In this chapter the adaptive response to radiation is examined by giving examples of the studies now being published in the literature, and from personal communications with leading investigators.

In these studies it has become apparent that two definitions of adaptive response are present in the design of the experiments. One type of study deals with adaptations that occur during the life span of a cell, tissue, or organism that may improve its chances for survival or lengthen its life span. The second type of study is that of an adaptive response of a population of cells or organisms that appears to be a permanent genetic change that again improves the chances for survival in the presence of radiation exposure. This latter type of study has difficulty distinguishing between a response by *selection* of an existing resistant phenotype and the induction of a new genetic mutant.

It is perhaps fitting in the new dollar-hungry political climate of the '90s to examine these adaptive responses very carefully and to determine if they may have an impact on human risk mitigation from radiation, or if the causative agents can be harnessed as prophylactic or mitigating agents against occasional or chronic exposure to radiation.

RESULTS AND DISCUSSION

If organisms can respond to an external insult by changing their internal milieu in order to overcome the deleterious effects of the insult and subsequent insults, it would seem logical that organisms should have a "trigger" for the response that is set at a very low dose. It is also pertinent to ask if the "triggering level" of the insult is normally part of the homeostatic environment.

These possibilities have been examined by investigators such as Planel[1-3] and coworkers, who have studied the effect of the natural background radiation on organisms. Dr. Planel, in particular, is aware of skepticism leveled at earlier experiments with small doses reporting radiation hormesis,[4-5] and thus has performed a series of elegant and very carefully controlled experiments examining the effect of removing the natural background radiation. An example of his results is shown in Figure 16.1. In this study, *Paramecium tetraurelia* were either shielded from background radiation, exposed to it, or shielded and exposed to Thorium equivalent to background. Dr. Planel clearly demonstrates that removal of background causes a decrease in population doublings. From these and many other experiments in single-celled organisms of both prokaryotic and eukaryotic origins, he has concluded that natural background or perhaps a little more radiation exposure is a part of the cells' homeostatic environment.

As we shall see later, irradiation-inducible proteins can be triggered by very small doses and can appear very rapidly. Thus, it is tempting to speculate that cells have evolved a complex response mechanism to external insults, and that natural radiation (perhaps among other agents) serves to keep the cell at a "surveillance" level of response.

If the response of cells to radiation is to produce proteins involved in protecting from and/or repairing DNA damage, one might expect to see an "adaptive" response in a cell population chronically exposed to small radiation doses that are above background.

However, since it is very well known that 'larger' doses of radiation are damaging, one has to ask the question, if a cell population can 'adapt' to

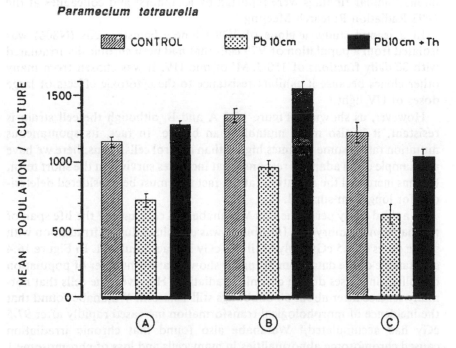

Figure 16.1. The growth rate is restored when shielded cultures are exposed to a level of ambient radiation comparable to background. Reproduced with permission from H. Planel et al., *Health Physics*, 52, 1987.

increase the likelihood of short-term survival, is this done at the expense of long-term deleterious effects in some of the population?

I will give three examples of experiments that perhaps address the question to some degree. First, a study in our laboratories where actively growing C3H10T1/2 cells were exposed to 10 cGy/day for 45 days, examined the survival and neoplastic transformation during the exposures and during subsequent acute exposures of gamma radiation.[6]

During the exposures there was no detectable cell killing and only a small induction of neoplastic transformation, leading us to the conclusion that at very low dose rates repair processes effectively remove or nullify most of the radiation damage.[6] However, as we present in Figure 16.2, when the chronically irradiated cell population was challenged with larger doses of high dose rate gamma radiation, the cells were more resistant to the cytotoxic effect of the radiation (upper curve, Panel A) and fewer transformants were induced compared to control cells given the same irradiations (lower curve, Panel B). While this result appears to be a positive adaptive response, it should be remembered that the cells have not been examined for chromosomal damage nor have they been cultured for extended periods, and that addition of a tumor promoter, TPA, to the cells caused the expression of a larger number of apparently latent gamma-ray-induced transformants. Similar findings were reported by Raaphorst and colleagues at the 1993 Radiation Research Meeting.[7]

In a second study, a clone of V79 Chinese hamster cells (N806) was isolated from a population of V79 cells that had been chronically irradiated with 80 daily fractions of 150-J/M² of mid-UV. It was chosen from many other clones because it exhibits resistance to the cytotoxic effects of large doses of UV light.[8]

However, as shown in Figure 16.3 A and B, although the cell strain is resistant, it is also more mutable than before. In fact, its spontaneous mutation rate is some 10 times higher than control cells. Thus, here we have an example of an adaptive response that increases survival in the short term, but has increased the mutation rate, a fact that must be considered deleterious for long-term survival.

In a third study performed by Watanabe and colleagues,[9] the life span of normal human embryonic fibroblasts was examined during irradiation with single doses of 7.5 cGy daily until 195 cGy had accumulated. In Figure 16.4 an example of the data from his paper shows that the number of population doublings increases during chronic irradiation. However, the cells that normally senesce after about 50 doublings still senesced. Watanabe found that the incidence of morphological transformation increased rapidly after 97.5 cGy had accumulated.[9] Watanabe also found that chronic irradiation caused chromosome abnormalities in many cells and loss of chromosome 1 in 70% to 88% of cells. Dr. Watanabe concluded from these studies that although chronic irradiation might increase the growth potential of cells,

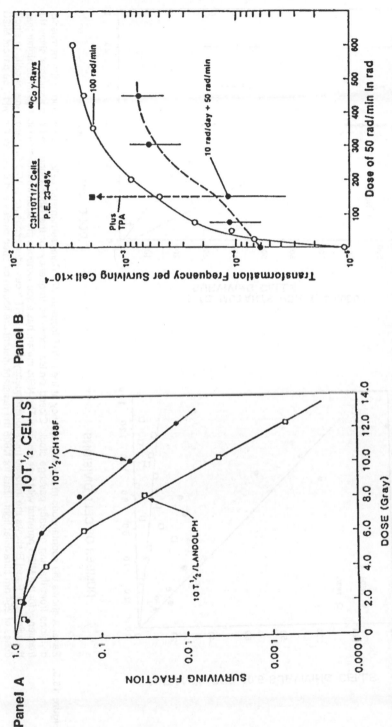

Figure 16.2.   Panel A: Surviving fraction of C3H10T1/2 cells after exposure to single acute doses of 50 kVp x-rays (open squares) or prolonged low dose rate gamma rays (300 cGy at 10 cGy/day) followed by single acute doses of 50 kVp X rays (closed circles). Reproduced with permission from C.K. Hill et al. *Radiat. Res.*, 119, 1989. Panel B: Transformation frequency in C3H10T1/2 cells after exposure to single acute exposures of gamma rays (100cGy/min) (open circles) or prolonged low dose rate gamma rays (300 cGy at 10 cGy/day) followed by single acute exposures of gamma rays (50 cGy/min) (closed circles). The closed square shows the data for the 150 cGy datum point (closed circle) when it was followed by weekly treatment with the promoter TPA at 1 µg/mL in the medium.

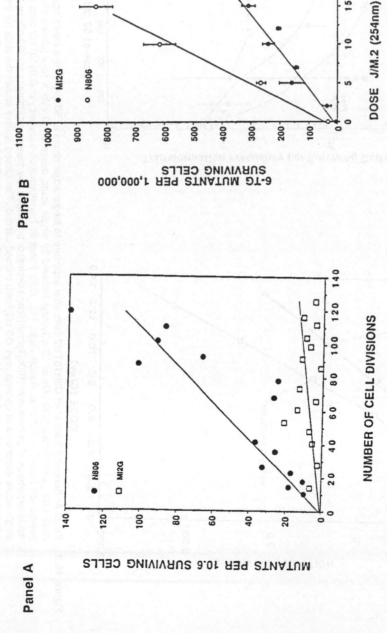

Figure 16.3. Panel A shows the spontaneous mutation frequency at the HPRT locus in V79 Chinese hamster cells as a function of the number of divisions from thawing of early passage stock. Closed circles show the frequency for the N806 clone, and open squares show the frequency for the MI2G parental wild type cell line. The N806 mutant has a spontaneous mutation frequency almost 10 times higher than the wild type. Panel B shows the induced mutation frequency at the HPRT locus after exposure of V79 Chinese hamster cells to single doses of 254 nm ultraviolet light. Data for N806 the resistant mutant are shown by open circles and those for the wild type parent MI2G are shown by closed circles.

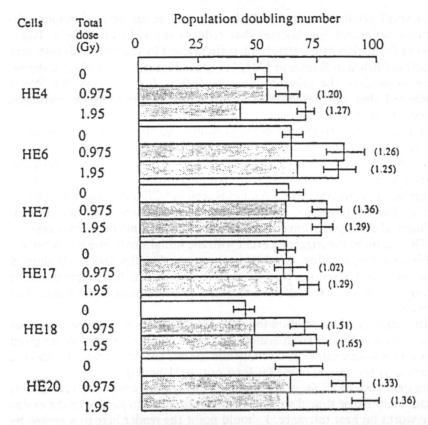

Figure 16.4. Population doubling number of human embryonic fibroblasts irradiated with protracted doses of ^{137}Cs gamma-rays (7.5 cGy/day). The number in parentheses shows the relative life span of irradiated cells to control cells. Each box with shadow shows the population doubling numbers of human cells irradiated with a single dose equivalent to the total protracted dose shown. Each bar shows the standard deviation of three independent determinations. Reproduced with permission from M. Watanabe, *Int. J. Radiat. Biol.*, 62, 1992.

this is not necessarily a positive adaptive response as the same cells had a cumulative burden of chromosomal changes.

All the studies mentioned so far have been observational and have not looked for mechanisms nor addressed the level of dose required to trigger a response. In the following series of experiments that began with studies of micronuclei and sister chromatid exchanges (SCEs), we will see the first evidence for induction of protein (some of which may be novel) by very small doses of radiation. It is these proteins that are presumed to be involved in repair or regulatory functions and whose induction may prevent or remove some of the damage from subsequent larger doses of radiation.

Several studies from Dr. S. Wolff's laboratory[10-14] have shown clearly how small priming doses of radiation can induce an adaptive response in human peripheral lymphocytes that reduces the yield of SCEs to larger doses of radiation given later. In their first report by Olivieri, Bodycote, and Wolff published in *Science* in 1984,[10] they used incorporated tritiated thymidine to produce the priming dose and followed with x-rays. Dr. Wolff suggested that this response was analogous to the adaptive responses reported for alkylating agents where a long low-level exposure reduces the damage occurring from large doses of similar agents given for a short time.

This work was independently confirmed by a report from Dr. Ikushima in a study using V79 Chinese hamster cells.[15] In Figure 16.5 the adaptive response after a priming dose from tritiated thymidine cannot be seen in micro nuclei induction (Panel A), but is seen in SCE induction (Panel B). In the studies by Ikushima, the follow-up radiation was from the thermal column of the Kyoto University Research Reactor and/or gamma rays.

The nature of the triggering event with the initial small dose has become a subject of close scrutiny. In subsequent reports, Shadley and Wolff showed that very small doses of x-rays given at acute dose rate (i.e., short exposure time) can also elicit the adaptive response to subsequent larger doses of x-rays.[13,16]

In summary, this radiation-induced adaptive response appears to occur after very small doses given acutely (0.25 to 2 cGy),[13] for small doses given at a very low dose rate (e.g. 0.01 Gy/min. or less for 0.5 Gy doses),[16] and the effect lasts for several hours (at least out to 11 hours).[10]

In deference to all the other papers, of which there are too many to list, about adaptive or should we say "stress" responses, in particular the extensive works on heat tolerance, I should point the reader here to a review by Dr. Adams et al.[17] given as a plenary lecture at The Kyoto Conference on low dose irradiation and biological defense mechanisms in August of 1992. In his review, Dr. Adams pointed out that many physical and chemical agents produce stress-responses. He noted that "the responses include the induction of resistance to higher doses of the same and other agents, increased rates of DNA replication, up-regulation of a variety of proteins, including some in cells signaling mechanisms with increased expression of some genes involved in cell growth and proliferation."

It is this generalized stress response and the proteins induced by the "stress" that the last part of this review will consider.

A good example of induced radiation resistance and cross resistance has been provided by the studies of Boreham et al. on the yeast *Saccharomyces cerevisiae*.[18] In Figure 16.6, I have reproduced some of their data showing how heat treatment, radiation treatment, and nutrient stress treatment can induce radiation resistance, and radiation treatment can induce thermal tolerance.

This type of cross-resistance adaptive response lends weight to the arguments already cited by Planel[1] and Adams,[17] and developed further in this chapter, that an inducible or an adaptive response may have evolved as a

Panel A

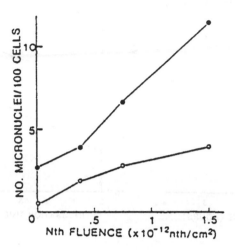

Panel B

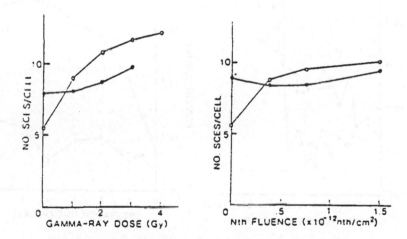

Figure 16.5.   Panel A: Effects of the ³H-dThD pretreatment on the induction of micronuclei by reactor radiations in Chinese hamster V79 cells. The pretreated (closed circles) and control (open circles) were irradiated with reactor radiations at the heavy water facility of KUR at 5 MW. Panel B: Effects of the ³H-dThd pretreatment on the induction of SCEs by the subsequent irradiations by gamma rays (left) and reactor radiations (right). The pretreated cells are shown by closed circles and the control cells by open circles. Panel A and B reproduced with permission from T. Ikushima, *Mutation Research*, 180, 1987.

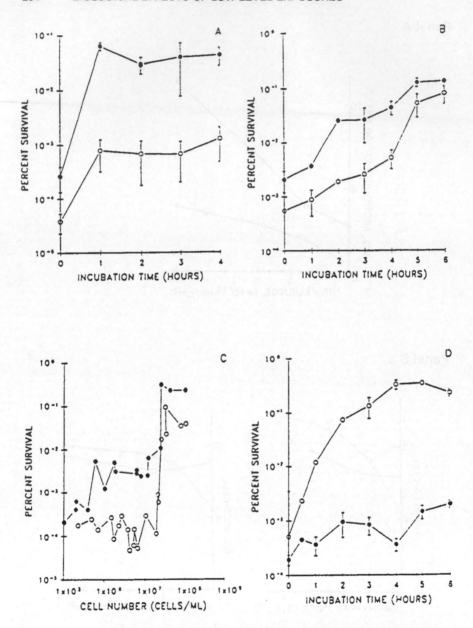

Figure 16.6. A. Heat-Induced radiation resistance. B. Radiation-induced radiation resistance. C. Nutrient stress-induced radiation resistance. D. Radiation-induced thermal tolerance. To induce resistance, wild type (open circles) and hsp104 mutant (closed circles) yeast cells were: given a 37° heat shock for 0–4 hours (A), given a 250 Gy gamma-ray inducing dose (B and D), or grown to different cell concentrations (C). Resistance was determined by measuring survival to a 1.75 kGy gamma-ray test dose (A, B, C) or a 4 minute 52°C heat shock (D). Reproduced with permission from D. R. Boreham et al. Personal communication, 1993.

response to occasional deleterious changes in the otherwise homeostatic environment of a species.

Now, although there have been many reports of radiation-induced proteins in the last decade (for reviews, see References 19–21), there have been few reports where investigators have looked explicitly for changes in gene expression or protein production after priming doses of radiation.

In the course of the studies of inducible proteins it has become clear, however, that some proteins are induced specifically by radiation, and not by other stress agents. For example, Boothman et al.[22] isolated a set of eight x-ray inducible proteins from certain human cells.

Recently, follow-up studies on inducible proteins and in cells where the survival or SCE adaptive response have been shown, have begun to suggest that protein changes can be detected after priming doses. I will give one example from the work of Boothman.[23] In Figure 16.7, the response of 4 of the x-ray inducible proteins are charted for a series of 5 cGy priming doses given at daily intervals, or for a single 450 cGy dose. Boothman notes that in the low dose primed HEp-2 and U1-Mel cells, the survival to a high challenging dose of irradiation was 50% to 100% higher. However, he also

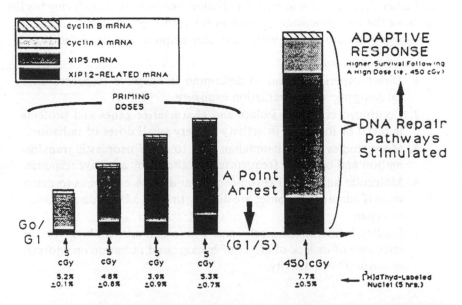

Figure 16.7. The bars show the increase in transcription levels for four inducible genes. On the left a gradual increase in transcript level is seen as 5cGy doses are repeated up to 4 times. On the right is shown the response for a single large dose of 450 cGy. Note that cyclin A is almost absent after the small priming doses (lightly stippled bar), but is present in significant amounts after the large dose. Reproduced with permission from Meyers et al. *Excerpta Medica, Low Dose Irradiation and Biological Defense Mechanisms,* Elsevier Sciences, 1992.

pointed out that in some normal human fibroblasts he did not see any change in survival. In the figure it is evident that expression of the genes is gradually increased by priming doses, and that the level of expression after priming doses is not necessarily the same after a high challenging dose. Boothman speculates that there may be feedback loops when cells are exposed to small repeated doses that maintain the optimum level of proteins required for cell repair and function.

So where does all the evidence presented here, and in the many papers I have no room to cite, leave us in regard to the role of adaptive responses in low level exposure.

Last year in my summary of molecular and cellular studies at the Kyoto conference[24] I stated strongly that the "No threshold theory of radiation *action*" should not be scrapped. I still stand by that assertion. The operative word is '**action**' instead of damage. All of the data show radiation causes a response in cells. The evidence to date suggests that cells and possibly organisms have evolved mechanisms allowing them to adapt to short-term insults to ensure survival, but this occurs at the possible expense of a higher long-term deleterious genetic effect. It is not yet clear what relevance, if any, these studies will have on the determination of risk estimates for cancer and other diseases from low-level radiation exposure in man during his life span, or the risk genetically to man in his future generations.

More studies are needed with particular emphasis on the following:

1. Direct experimentation to determine if adaptive responses alter carcinogenic risk of radiation exposure.
2. Continued efforts to isolate and characterize genes and proteins induced or increased in activity by very small doses of radiation.
3. More studies using mammalian cells to see if neoplastic transformation and mutation frequencies are altered by adaptive response.
4. Molecular analyses of chromosomal and DNA endpoints to determine if adaptive responses are directly linked to DNA damage and/or repair.
5. Studies in mammalian cells and whole animals, looking for the influence of or lack of effect of background radiation on "normal or natural" cell activity.

ACKNOWLEDGMENTS

I would like to thank all the authors who kindly sent me reprints of their studies, suggestions for this chapter, and permission to quote their work. I hope I have done them justice.

REFERENCES

1. Planel, H., J.P. Soleilhavoup, R. Tiaxzdor, G. Richoilley, A. Conter, F. Croute, C. Caratero, and Y. Gaubin. "Influence on Cell Proliferation of Background Radiation or Exposure to Very Low, Chronic Gamma Radiation," *Health Physics*, 52:571 (1987).

2. Conter, A., D. Dupouy, and H. Planel. "Demonstration of a Biological Effect of Natural Ionizing Radiations," *Int. J. Radiat. Biol.*, 43:421 (1983).

3. Planel, H., J.P. Soleilhavoup, and R. Tixador. "Recherches sur l'action dea radiations ionisantes naturelles sur la croissance d'etre unicellulaires," *Compte-Rendus Acad Sci.*, 260:3770 (1965).

4. Sidraki, G.H. "Effects of Low Doses of Ionizing Radiation on the Growth and Respiration of Yeast and Broad Beans," *Stimulation Newsletter*, 1:1 (1970).

5. Conter, A., D. Dupouy, and H. Planel. "Effects of Dose Rate on Response of *Synechococcus lividus* to Very Low Doses of Chronic Gamma Radiation: Influence of Enzymatic Equipment of Starting Cells," *Radiat. Research*, 105:379 (1986).

6. Hill, C.K., A. Han, and M.M. Elkind. "Promoter-Enhanced Neoplastic Transformation at 10 cGy/day," *Radiat. Research*, 119:348 (1989).

7. Raaphorst, G.P., and R.E.J. Mitchel, "Radiation-Induced Adaptive Response to Ionizing Radiation in C3H 10T1/2 Cells," in Abstracts, Forty-first annual meeting of the Radiation Research Society, Dallas, TX, March 20–25, 1993, p. 84.

8. Ikebuchi, M., M. Osmak, A. Han, and C.K. Hill. "Multiple, Small Exposures of Far-Ultraviolet or Mid-Ultraviolet Light Change the Sensitivity to Acute Ultraviolet Exposure Measured by Cell Lethality and Mutagenesis in V79 Chinese Hamster Cells," *Radiat. Research,* 114:248 (1988).

9. Watanabe, M., M. Suzuki, K. Suzuki, K. Nakano, and K. Watanabe. "Effect of Multiple Irradiation with Low Doses of Gamma-Rays on Morphological Transformation and Growth Ability of Human Embryo Cells *in Vitro*," *Int. J. Radiat. Biol.*, 62:711 (1992).

10. Olivieri, G., J. Bodycote, and S. Wolff. "Adaptive Response of Human Lymphocytes to Low Concentrations of Radioactive Thymidine," *Science*, 223:594 (1984).

11. Wienke, J.K., V. Afzal, G. Oliviera, and S. Wolff. "Evidence That the [³H-Thymidine-Induced Adaptive Response of Human Lymphocytes to Subsequent Doses of X-Rays Involves the Induction of a Chromosomal Repair Mechanism," *Mutagenesis*, 1:375 (1986).

12. Wolff, S., R. Jostes, F.T. Cross, T.E. Hui, V. Afzal, and J. Wienke. "Adaptive Response of Human Lymphocytes for the Repair of Radon-Induced Chromosomal Damage," *Mutation Research*, 250:209 (1991).

13. Shadley, J.D., Z. Afzal, and S. Wolff. "Characterization of the Adaptive Response to Ionizing Radiation Induced by Low Doses of X-Rays to Human Lymphocytes," *Radiat. Research*, 111:511 (1987).

14. Wolff, S., Failla Memorial Lecture, "Is Radiation All Bad? The Search for Adaptation," *Radiat. Research*, 131:117 (1992).

15. Ikushima, T. "Chromosomal Responses to Ionizing Radiation Reminiscent of an Adaptive Response in Cultured Chinese Hamster Cells," *Mutation Research*, 180:215 (1987).

16. Shadley, J.D., and J.K. Wiencke. "Induction of the Adaptive Response by X-Ray Is Dependent on Radiation Intensity," *Int. J. Radiat. Biol.* 56:107 (1989).

17. Adams, G., W-S. Chan, and I.J. Stratford. "Radiation and Other Stress Responses and Their Relevance to Cancer: Some General Features," in *Proceedings of the International Meeting on Low Dose and Biological Defense Mechanisms*, T. Sugahara, L.A. Sagan, and T. Aoyama, Eds., Kyoto, August 1992, pp. 389. Elsevier Science Publishers, 1993.

18. Boreham, D.R., M.E. Bahen, S. Laffrenier, and R.E.J. Mitchel. "Stress-Induced Radiation Resistance in Yeast," *Radiat. Res.*, 137(2): (1994).

19. Walker, G.C. "Inducible DNA Repair Systems," *Ann. Rev. Biochem.*, 54:425 (1985).

20. Friedberg, E.C., *DNA Repair*, (New York: W. H. Freeman Co., 1985), p. 25.

21. Fornace, A.J., Jr., I. Alamo, Jr., and M.C. Hollander. "DNA Damages Inducible Transcripts in Mammalian Cells," *Proc. Natl. Acad. Sci. USA*, 85:8800 (1988).

22. Boothman, D., I. Bouvard, and E.N. Hughes. "Identification and Characterization of X-Ray-Induced Proteins in Human Cells," *Cancer Res.* 49: 2871 (1989).

23. Meyers, M., R.A. Schea, A.E. Petrowski, H. Seabury, P.W. McLaughlin, I. Lee, S.W. Lee, and D.A. Boothman. "Role of X-Ray-Inducible Genes and Proteins in Adaptive Survival Responses," in *Proceedings of the International Meeting on Low Irradiation and Biological Defense Mechanisms*. T. Sugahara, L.A. Sagan and T. Aoyama, Eds., Kyoto, August, 1992, p. 263, Elsevier Science Publishers, 1993.

24. Hill, C.K. "Molecular and Cellular Level Studies," in *Proceedings of the International Meeting on Low Dose Irradiation and Biological Defense Mechanisms*, T. Sugahara, L.A. Sagan, and T. Aoyama, Eds., Kyoto, August, 1992, p. 469, Elsevier Science Publishers, 1993.

PART IV

BELLE Conference Perspectives and Summary

CHAPTER 17

Synopsis of the BELLE Conference on Chemicals and Radiation

John D. Graham, School of Public Health,
Harvard University, Center for Risk Analysis, Boston, Massachusetts

Despite the diversity of presentations and issues covered in this conference, I detected a dominant theme in the discussions: a search for a new scientific paradigm to inform dose-response evaluations of chemicals and radiation. The beginnings of a new paradigm seem to be emerging in the notion of a U-shaped dose-response curve, but the vision was sometimes cloudy and confusing. As a relative newcomer to the field, here are some critical thoughts about what has been said at the conference and some suggestions about what might be done to nurture the alternative paradigm.

Dr. Schaffner described the prevailing paradigm, which presumes that any agent shown to be harmful at high doses will be harmful at low doses. This "no-threshold" paradigm appears to have had its origins in the study of radiation and human cancer, was later extended to the study of chemicals and cancer, and has been proposed to be applicable to some noncancer effects such as the neurobehavioral effects of lead.

As Dr. Sagan has indicated, the prevailing paradigm is supported by both scientific speculation and social philosophy. The scientific underpinnings are traced to early models of carcinogenesis such as the one-hit model and its more refined variants. The notion of wide variation in human susceptibilities to adverse health effects has also played a role in sustaining the notion of low-dose, adverse health effects. In the arena of social policy, the principal argument has been "better safe than sorry." When in doubt about the health effects of low doses, we should assume that they exist and that *they are adverse*! This policy position is sometimes stated by scientists and government officials as if it were scientific fact, even though it has not been established with certainty. It is of course the politically correct position these days, and we all recognize that the scientific community is not immune from the ills of political correctness.

The societal ramifications of the prevailing no-threshold paradigm, correct or incorrect, are profound. There has been an explosion of public and private spending aimed at reducing human exposures to low doses of chemicals and radiation. This industry is now growing at a rate that is roughly twice the rate of the U.S. gross national product. Concern about low-dose effects is also influencing the course of innovation in the chemical and pharmaceutical industries. Documentation of high-dose health effects can result in the termination of programs to develop new products, chemicals, pesticides, and drugs. The threat of low-dose effects is also fueling new social policies such as "Toxics Use Reduction" (less of a bad thing must be good) and zero discharge of "toxics" into the Great Lakes and other bodies. Various "right to know" laws are also based on the premise that citizens need to be informed about low-dose exposures to chemicals and radiation.

In the course of this conference, I was struck by the subtle yet powerful influence that the current paradigm is exerting on the course of scientific inquiry and experimentation. Much of the trend is toward designing studies that will not be capable of revealing any protective effects that may exist at lower doses. As Dr. Calabrese indicated, there is the preoccupation with establishing lower and lower no adverse effect levels (NOAELs) of exposure, without considering the severity or frequency of these effects in the exposed population or the possibility of beneficial effects for less sensitive endpoints and populations. Nor does the testing of chemicals at the maximum tolerated dose (MTD) permit us to examine the possibility of "paradoxical" dose-response curves at low doses, such as those suggested in Dr. Cook's presentation. In epidemiology studies, the blind use of duration of exposure as a surrogate for dose may mix the adverse effects of high doses with the hypothesized beneficial effects of low doses, which can be expected to produce the ambiguity that plagues most epidemiological studies. The way animals are fed in laboratory studies creates such variability problems that any subtle protective effects at the lower doses may be too difficult to detect, even though the blunt adverse effects at the MTD are discernible. We should also remember that negative results are generally less publishable than positives unless the negative result has been preceded by a string of positives (in which case the negatives may be discarded). Since protective effects are to be found in the negative results (assuming they exist), publication bias is a very salient issue. This pattern also means that there will be fewer opportunities to find published replications of negative results, which might otherwise induce us to seriously consider the possibility of protective effects.

Despite the formidable biases against finding protective effects at lower doses, there are certain "anomalies" that seem to be incompatible with the prevailing paradigm. According to Dr. Schaffner, such anomalies were highlighted in Kuhn's *Structure of Scientific Revolutions* as a forerunner of shifts in fundamental thinking. For instance, how does the current paradigm accommodate growing evidence that moderate alcohol consumption is protective against heart disease even though heavy consumption is pathological? How does the current paradigm accommodate the findings reported by Drs. Hart and Turturro that both too little and too much caloric intake reduce longevity in rodents? What about the evidence that low doses of aspirin may be protective against heart disease even though aspirin is indisputably toxic at high doses? From my perspective, the most fascinating anomaly was the data presented by Drs. Cook and Gallo, which suggest that while dioxin causes liver tumors in rodents at high doses it may act to prevent breast cancer at low doses. In assessing these facts, the question becomes: Are these anomalies or do they represent the emergence of a refined or new paradigm?

In the spirit of supporting the development of a new perspective, here are some suggestions that might help promote the "U-shaped" paradigm or its variants.

First, history suggests that advocates of any new paradigm can be harmed by the use of flimsy or premature empirical arguments. In Dr. Calabrese's historical review, he found that many of the early claims on behalf of hormesis were based on poor reasoning. In occupational studies, for example, the well-known healthy-worker effect may be mistaken for a protective effect. In animal tests, subtle differences in caloric intake among rodents may act as a powerful confounder that creates the false impression that low doses are protective. My point is that the credibility of the U-shaped paradigm can be hurt by advocates who make weak empirical arguments on its behalf.

Second, and more constructively, there is an urgent need for the development, publication, and communication of several well-documented case studies of the U-shaped dose response curve. While many toxicologists will claim that such knowledge has existed for decades (or even centuries), this information is not well known among current opinion leaders, policymakers, advocates, and scientists. For if such evidence is valid, reducing low doses of chemicals and radiation will not satisfy our desire to be safe rather than sorry!

Third, advocates of the U-shaped paradigm or its variants need to move beyond empiricism to explanation. If we can explain why low doses are protective (not just why they seem to be), the prospects of a genuine scientific revolution are much greater. In this regard, I was intrigued by the mechanistic studies described in the conference by Drs. Mehendale, Smith-Sonneborn, and Hill. We need to know much more about the potential protective effects of growth stimulation, responses to stress, promotion of tissue repair, and other biological processes.

Finally, and most importantly, we need to be acutely sensitive to the construction of asymmetric burdens of proof in the scientific literature and the process of consensus formation. The papers by Drs. Schaffner and Sagan warned us of the generic danger: proponents of a prevailing paradigm will tend to ignore, suppress, and censor evidence that seems to be incompatible with or threatening to the status quo. I would like to take the

liberty of picking on the fine paper prepared by Dr. Gaylor for this conference, since I believe it may contain some subtle illustrations of asymmetric
burdens of proof. Consider the following passages from Dr. Gaylor's paper:

- p. 5: "A statistically significant reduction in toxicity at a low dose
  compared to a control group is not a sufficient demonstration of
  hormesis."
- p. 6: "A high level of statistical significance, e.g., P less than 0.01,
  should probably be required to avoid concluding that hormesis
  exists when it doesn't exist."
- p. 6: "A good statistical fit of a hormetic dose-response curve to
  bioassay data does not necessarily constitute proof of hormesis."
- p. 7: "Even though no single set of data may provide convincing
  evidence of hormesis, consistency of effects in two or more studies
  may provide adequate evidence for hormesis."

Now, Dr. Gaylor's observations may indeed reflect rigorous scientific
standards, but my point is that these burdens of proof seem somewhat more
exacting than is typically required to incriminate an agent for its toxicity,
particularly if it is to be considered a "carcinogen." For example, I urge
readers to compare Dr. Gaylor's proof burdens for hormesis to the proof
burdens used in, say, interpretation of results from bioassays sponsored by
the National Toxicology Program.

Now that I have picked on Dr. Gaylor's paper I would like to conclude by
praising it, since it advances what I believe is the most critical concept in the
conference: the "optimum dose." When background doses (from all
sources) exceed the optimum dose, the hormesis concept is irrelevant to
social policy. When the optimum dose exceeds the background dose (from
all sources), the hormesis hypothesis has profound implications for social
policy. By implicitly rejecting the notion that zero dose is always the optimum dose, Dr. Gaylor's paper opens up a vast array of scientific inquiries
that promise to challenge the prevailing paradigm.

Thank you very much for the opportunity to offer these somewhat
extemporaneous remarks.

CHAPTER 18

New Perspectives on Dose Response Relationships and Low Level Exposures

John Higginson, Department of Community Medicine,
Georgetown University Medical Center, Washington, DC

The discussion today emphasizes that health risks, especially those due to cancer, are an important aspect and driving force in society's approach to ecological and environmental issues. Since any expenditure in environmental health should be considered as part of the national health dollar, prudent social management indicates the necessity to develop methodologies to identify and distinguish trivial (or nonexistent) risks from significant hazards. Such a knowledge is essential to establish appropriate public health priorities.

It is clear from the discussion today that existing methodologies are based on models containing theoretical assumptions for which the scientific foundation remains limited and usually has not been validated. Much of the controversy relates to the calculated theoretical risks in humans associated with low-level environmental exposures to potential carcinogens. Such levels of exposure are far below those at which an effect can be detected and consequently used in validating such estimates in humans. Thus, managerial decisions on such risks remain, largely, judgmental. This uncertainty ensures that they will be subjected to a range of external criticisms from politicians, activists, academics, industrialists, and other special interest groups. It significantly molds perceptions among the public as to the facts.

This problem of the evaluation of low risks exists in both North America and Europe. However, there still remains in Europe a greater willingness for society, and legislative bodies, to accept as credible the views of public officials and their scientific advisors. Despite an ever-growing green movement, managerial decisions remain more pragmatic than in the U.S. In contrast, in North America, mistrust in public servants and scientists tends to be common and it is much more difficult for the public to accept that an objective and knowledgeable consensus can exist among technical experts. Further, divergent views are regarded as of equal validity irrespective of their scientific base and support, and often the discussion enters the category of a high school football match.

Based on earlier experience with radiation, this has led to attempts in the United States to develop biomathematical models for measuring cancer risks numerically based on simplistic extrapolatory models on the grounds of objectivity [Quantitative Risk Assessment (QRA)]. Such models contain certain assumptions that cannot be validated in humans, and they are usually expressed as upper bounds linear extrapolations based on the worst case scenario. There appears to be a general consensus at the meeting as to their scientific and biological deficiencies.

It is unnecessary to go through the historical events and repeat the discussions that have been adequately covered by others but it is of interest to note that for practical purposes no significant new carcinogens in humans have been identified since the early 1970s, although a number of animal carcinogens have been alleged to cause problems. This contrasts with the situation, over 40 years ago, when there was a wide acceptance among many scientists that general pollution due to industrial products could be a major factor in the causation of lung and other cancers outside the occupational setting. As is well known, events over the later years have not confirmed this view, with the exception of asbestos. However, the public perceptions as to the danger of xenobiotics, if man-made, are now so strongly embedded that managerial and political decisions must take these into account.

It is probably true that we know enough to protect society with reasonable efficiency at the present time. Further, the methodological techniques now available to screen new compounds in animals and in vitro systems make it unlikely, though not impossible, that any new *major* carcinogen will enter the environment. Nevertheless, for the moment it is probable that animal data, despite its limitations, will continue to be used for the estimation of low level dose responses. To avoid bureaucratic sclerosis, the limitations of such dose estimates should be continually examined. Unfortunately, more epidemiological studies, irrespective of their low exposures, are most unlikely to produce data that can confirm or disprove the animal estimates, and, further, also carry the danger of demonstrating random and noncausal association when done on a small scale.

There is considerable hope that the inclusion of mechanistic data such as pharmacokinetics may significantly improve assumptions in QRA, but this remains in the future. On the other hand, knowledge of molecular mechanisms may greatly enhance scientific judgments based on a "weight of the evidence." Thus, for example, they may show important biological differences between animal and human cells and may provide more information on how molecular mechanisms may work. Nonetheless, models, for the moment, cannot substitute uncritically for scientific judgment.

In conclusion, the question still remains as to whether target cell reactions are qualitatively different at low, as compared to high level exposures. Several speakers at this conference have discussed the issue of *hormesis*. It is clear that there is no definite evidence to exclude the possibility that mechanisms at low level exposure may not be qualitatively different from high exposures. The discussion of hormetic studies are tantalizing in their implication as to thresholds, etc., in considering such issues. However, I believe from a political viewpoint in trying to rewrite the Delaney Clause, it is doubtful if, at this time, such studies will have much impact.

The difficulties of QRA have been emphasized by earlier speakers and it seems undesirable to continue to rely so heavily on methods which have so many fallacies. The expression that we acquire more "good science" is often used. However, while sound mathematical theory can be good science, it does not necessarily reflect reality. In fact, what is meant is better and more objective scientific judgment, through the weight of the evidence approach. This implies subjectivity, and may not be acceptable to the public.

In the meantime, we are left with the problem of distinguishing the trivial or nonexistent risk from a major risk. If models are not to be used in regulation, it will be necessary to offer some generally acceptable alternative. The only valid alternative I see at present is the use of a weight of the evidence approach based on the totality of the data including molecular mechanisms.

CHAPTER 19

BELLE Conference: A Summary

Leonard Sagan, Electric Power Research Institute,
Palo Alto, California

Approximately 100 participants attended the second annual BELLE Conference held in Crystal City, Virginia on April 26, 27, 1993. Greetings were expressed by both Edward Calabrese of the University of Massachusetts, who is also Director of BELLE, and Arthur Wykes, President of the Association of Government Toxicologists. Calabrese delivered a review of the development and goals of BELLE. He emphasized the need to "Let the data lead us" without imposing constraints on interpretation.

There were two opening statements on the general subject of scientific paradigms. Kenneth Schaffner, a physician and philosopher of science at George Washington University, discussed the concept of scientific paradigms, as developed by Thomas Kuhn in his 1970 book, *The Structure of Scientific Revolutions*. Kuhn had suggested that science operates on models ("paradigms") which are revised only when contrary information becomes overwhelming and a "scientific revolution" occurs. Schaffner discussed some developments in Kuhn's concept of paradigms and their relevance to the BELLE meeting.

In his presentation, Leonard Sagan (Electric Power Research Institute) turned attention to the radiation paradigm, its history, and the assumptions inherent in that paradigm. The radiation paradigm, which has subsequently been expanded to include effects of exposures to chemicals at low doses, implies the absence of a threshold and deleterious (but nondetectable) effects at very low exposures. Sagan suggested that the development of this model was very much influenced by the environmental ethic of the 1950s and 1960s. He also suggested that there are many sectors of society (e.g., lawyers, regulators, radiation scientists) who now have a stake in maintaining the model, even though there is little scientific validation, and growing economic costs and little health benefit.

George Milo (Ohio State University) discussed the inconsistencies/consistencies of expression of biological endpoints with the expression of molecular mutations when subacute nontoxic dosages of activated environmental xenobiotics activate oncogenes and/or other molecular mutations in human cells. He discussed the issue of whether humans fit the present linear animal model of progression used to explain malignant progression. The overinterpretation of animal data has not shed a great deal of light on the complexity of the biological carcinogenic endpoint when human cells were exposed to either subacute or chronic dosage levels of exposure to activated environmental xenobiotics.

Three speakers examined the statistical basis for the low-dose non-threshold model. Mike Davis (Environmental Protection Agency) reported a systematic review of a sample of 1,800 articles containing dose-response data culled from the toxicology literature. Applying certain criteria, 147 of these articles were selected for more detailed analysis. Of these, 22 (15%) showed evidence of a "U-shaped" dose-response relationship. Davis illustrated some of the conceptual and statistical difficulties in ascertaining the incidence of nonmonotonic relationships.

David Gaylor (National Center for Toxicological Research) suggested that there may be an optimal exposure level for all environmental agents. Hormetic responses will only appear, he said, if the optimal dose is greater than the naturally occurring environmental level. If the optimal level is less than the environmental level, then there will be no stimulatory or hormetic effect from increasing exposure levels. Gaylor also reported his own review of some 3,000 animal studies carried out for carcinogenesis. In none of them did he find evidence of hormesis, although he admitted that these animal studies, carried out at relatively high doses and with relatively few animals, are less than ideal for the examination of a beneficial effect.

Peter Groer (University of Tennessee) described Bayesian techniques for examining low dose data, searching for "change points" in the data. In both beagle and human data, he demonstrated evidence of such change points. The human data were derived from studies of women watch dial painters with body burdens of radium 226, and with bone cancer as an endpoint.

Four speakers addressed the question of mechanisms: how could stimulatory effects at low doses be explained? Are observations of phenomena at low doses consistent with the linear paradigm? Harihara Mehendale (Northeast Louisiana State University) sees stimulated tissue repair as central to an understanding of the response to a toxic agent. If the initial exposure is large, then the repair mechanism is unable to respond, whereas with smaller doses the repair response appears and becomes instrumental in recovery. Exposure to a second inhibitory agent or event can also interfere with repair, resulting in much higher toxicity.,

Colin Hill (University of Southern California) spoke of a class of phenomena becoming known as an "adaptive response." There are two types of adaptive response that appear to be elicited by radiation. One is a response during a cell's life cycle, usually inducible, that reduces the toxic effects of radiation insult, thus resulting in more cells surviving. The other is a heritable adaptive response that can be passed on from one generation to another.

Again, the end result is, at least for a short term, an increase in survival. However, Hill showed that, at least in the later type of adaptive response, the long-term effect of an inducible change may be genetically deleterious. Although adaptive responses are now considered real phenomenon, there is still some confusion as to their exact nature. He concluded that understanding of adaptive responses and their incorporation into risk assessment will only occur when some systematic use of models is adopted.

Joan Smith-Sonnenborn of the University of Wyoming focused on the role of stress proteins as pivotal in the adaptive response. She noted that different patterns of proteins are synthesized in response to a wide variety of stressors, and speculated that these proteins may play a role in a variety of protective mechanisms, including DNA repair and increased cell survival.

Angelo Turturro (National Center for Toxicological Research) described the series of experiments done at NCTR to investigate the mechanisms underlying the well-known increase in survival and decrease in carcinogenesis of caloric restricted animals. Many physiological parameters, including energy metabolism, circadian rhythms, and DNA repair, are altered in these animals. From these observations, it would appear that dietary factors may play an important role in modifying dose-response relationships, perhaps making them U-shaped.

Although unable to attend, James Trosko's (Michigan State University) submitted paper raised some interesting questions about the role of radiation in carcinogenesis. Within the context of the current paradigms of the multistage concept of carcinogenesis, stem-cell theory of cancer and the oncogene/tumor suppressor gene theory, the fact that ionizing radiation induces mostly deletion mutations and chromosomal rearrangements calls into question acute, low dose exposures of radiation as a "complete carcinogen" and nonthreshold model. He noted that the initiation/promotion/progression stages of carcinogenesis involve multiple genetic and epigenetic events, and are mechanistically different. Different mechanisms can lead to the initiation, promotion, or clonal expansion of a stem cell and its ultimate conversion to a metastatic cell. Therefore, Trosko pointed out that while radiation is associated with the appearance of cancers after exposure, several questions have to be answered, such as: "Does ionizing radiation affect all or just some of these stages? Does ionizing radiation only contribute by mutagenesis, or can it contribute to promotion by cytotoxicity-induced hyperplasia and to the initiation/progression stages by epigenetic alteration of gene expression?"

Three epidemiologists presented views on human responses to low doses of chemical agents and radioactivity. Ethel Gilbert (Pacific Northwest Laboratory) reported on her studies of workers exposed to occupational levels of radiation. In her view, epidemiological studies do not have sufficient precision to resolve the issue of low dose effects, whether harmful or protective. The principal use of these studies will be to provide an upper bound on risks.

Robert Miller (National Cancer Institute), from a review of the literature, found that the lowest doses at which effects of radiation have been demonstrated are: 0.04 Gy for thyroid cancer among Israeli children given radiotherapy for ringworm of the scalp; 0.2 Gy for somatic cell mutations detectable by the glycophorin A test; 0.1 to 0.0.2 Gy for small head size after in utero exposure to the atomic bomb at 4–17 wk of gestational age; 0.2 Gy for long-lasting chromosomal aberrations; 0.2–0.5 Gy for leukemia and breast cancer among Japanese A-bomb survivors, with the highest relative risk among those exposed under 15 years of age; and 0.6 Gy for severe mental retardation, with brain damage found through Medical Res Image after in utero exposure to radiation from the atomic bomb at 8–15 wk of gestational age. Diagnostic x-ray exposures later in pregnancy have been related to a 1.5-fold excess of childhood leukemia in Great Britain and New England.

As has been known for some time, cancer has been noted in occupationally exposed human populations working with chemicals. In some cases however, as pointed out by Ralph Cook (Dow Corning), there is evidence of a protective effect. This was illustrated with data from dioxin exposed populations, an observation which is consistent with laboratory animal data.

Several speakers were asked to comment on the proceedings of the meeting. One of these was John Graham (Harvard School of Public Health). Graham noted that during the two-day symposium, he heard several examples of agents which appeared to be inconsistent with the nonthreshold paradigm. They were: alcohol, diet, dioxin, and aspirin, all of which produced paradoxical (and beneficial) effects at low doses. He suggested that scientists consider several tactics for achieving a new paradigm. The first was "Don't over-reach," by which he meant that scientists should be cautious about generalizing from weak data. Secondly, scientists should be open to exceptions to rules, and should be bold in disseminating those exceptions. Thirdly, we must continue to conduct the mechanistic studies necessary to build a base of knowledge about low dose effects. Finally, he advised that there be a symmetry in our standards of proof, being just as tough on those observations which appear to prove the rule (e.g., the nonthreshold model) as on those which seem to be exceptions to the rule (hormesis).

Both Roger McClellan (Chemical Industry Institute of Toxicology) and John Higginson (Georgetown University) expressed concern about the dogmatic application and misapplication of the nonthreshold model. Higginson proposes that rigid risk assessment, based on a linear model, be abandoned, and that a more pragmatic common sense judgmental approach be utilized based on a weight of the total evidence analysis, including mechanisms. McClellan closed the meeting, emphasizing the need to conduct studies of carcinogenesis at realistic levels of exposure, rather than at near toxic levels.

APPENDIX

BIOLOGICAL EFFECTS OF LOW LEVEL EXPOSURES

Where is BELLE and Who Sponsors it?

What is BELLE?
Where is BELLE and Who Sponsors it?

What is BELLE?

In May, 1990 a group of scientists representing several federal agencies, the International Society of Regulatory Toxicology and Pharmacology, the private sector, and academia met to develop a strategy to encourage the study of the biological effects of low level exposures (BELLE) to chemical agents and radioactivity. The meeting was convened because of the recognition that most human exposures to chemical and physical agents are at relatively low levels, yet most toxicological studies assessing potential human health effects involve exposures to quite high levels, often orders of magnitude greater than actual human exposures. Consequently, risks at low levels are estimated by various means, frequently utilizing assumptions about which there may be considerable uncertainty.

The BELLE Advisory Committee is committed to the enhanced understanding of low-dose responses of all types, whether of an expected nature (e.g., linear, sublinear) or of a so-called paradoxical nature. Paradoxical dose-response relationships might include U-shaped dose-response curves, hormesis, and in some restrictive sense, biphasic dose-response curves. Although there are many scattered reports of such paradoxical responses in the biomedical literature, these responses have not generally been rigorously assessed, nor have the underlying mechanisms been adequately identified. Laboratory and regulatory scientists have tended to dismiss these paradoxical responses as anomalies inconsistent with conventional scientific paradigms.

The focus of BELLE encompasses dose-response relationships to toxic agents, pharmaceuticals, and natural products over wide dosage ranges in in vitro systems and in vivo systems, including human populations. While a principal emphasis of BELLE is to promote the scientific understanding of low-level effects (especially seemingly paradoxical effects), the initial goal of BELLE is the scientific evaluation of the existing literature and of ways to improve research and assessment methods.

Where is BELLE and Who Sponsors it?

The BELLE program is directed by the BELLE Advisory Committee, which is listed in the front matter of this book. It is administratively managed by Edward J. Calabrese, Professor of Toxicology, School of Public Health, University of Massachusetts, Amherst. The BELLE Advisory Committee meets regularly and would encourage those interested in attending committee meetings and becoming involved in BELLE activities to contact the BELLE office or one of the Advisory Committee members.

The sponsorship of BELLE is intended to be broadly based, with financial and other support from both the public and private sectors. The finances are managed by the University of Massachusetts, School of Public Health.

Acronyms and Abbreviations

AAQS	ambient air quality standard
ACDQ	6-amino-7-chloro-5,8-dioxoquinoline
ACGIH	American Conference of Governmental Industrial Hygienists
AECL	Atomic Energy of Canada Limited
AEL	acceptable exposure level
AIG	anchorage independent growth
BEIR	Biological Effects of Ionizing Radiation
BELLE	biological effects of low level exposures
BOF	basic oxygen furnace
BW	body weight
CF	confounding factor
COH	coefficient of haze
CR	caloric restriction
DDREF	dose and dose rate effectiveness factor
DE	dose equivalent
DNA	deoxyribonucleic acid
d-r.c.s	dose-response curves
EPA	Environmental Protection Agency
EPCRKA	Emergency Planning and Community Right to Know Act
EPR	early phase regeneration
GJIC	gap junctional intercellular communication
gr/SDCF	grains per standard dry cubic foot
HAP	hazardous air pollutant
IARC	International Agency for Research on Cancer
ICE	internal combustion engine
ICRP	International Commission of Radiological Protection
IDLH	immediately dangerous to life and health
LDEF	low-dose extrapolation factor
LET	low energy transfer
LHC	lymphatic and hematopoietic cancer
MSA	metropolitan statistical area
MTD	maximum tolerated dose
NAAQS	national ambient air quality standards
NAMS	National Air Monitoring Station
NIOSH	National Institute for Occupational Safety and Health
NOAEL	no observed adverse effect level
NOEL	no observed effect level
NNW	non-nuclear workers
NRRW	National Registry for Radiation Workers
NTP	National Toxicology Program
OEL	occupational exposure level

ORNL	Oak Ridge National Laboratory
OSHA	Occupational Safety and Health Administration
PAN	peroxyacyl nitrate
PDL	population doubling
PEL	permissible exposure level
PICs	products of incomplete combustion
PIH	pregnancy induced hypertension
PM	particulate matter
PSD	Prevention of Significant Deterioration
PSI	pollutant standards index
QRA	quantitative risk assessment
RERF	Radiation Effects Research Foundation
RLI	radiolabeling index
RICE	reciprocating internal combustion engine
RR	relative risks
SCE	sister chromatid exchange
SD	standard deviation
SD	Sprague-Dawley
SEV	socioeconomic variable
SI	spark ignited
SICs	Standard Industrial Classifications
SLAMS	state and local air monitoring stations
SMR	standardized mortality rate
SPM	special purpose monitor
STEL	short-term exposure level
STS	soft tissue sarcoma
TAP	Technical Advisory Panel
TLV	threshold limit value
TWA	time-weighted average
UCF	unrecognized confounding factor
URF	unit risk factor
WLM	working-level-month
ZED	zero equivalent dose

List of Contributors

Edward J. Calabrese, School of Public Health (N344), University of Massachusetts, Amherst, MA 01003

Bruce A. Carnes, Biological and Medical Research Division, Argonne National Laboratory, Argonne, IL 60439

Bernard L. Cohen, Physics Department, University of Pittsburgh, Pittsburgh, PA 15260

Ralph R. Cook, Occupational Health and Epidemiology, Dow Corning Corporation, Mail #CO1120, Midland, MI 48686-0994

J. Michael Davis, Environmental Criteria and Assessment Office (MD-52), Office of Health and Environmental Assessment, U.S. Environmental Protection Agency, Research Triangle Park, NC 27711

David W. Gaylor, National Center for Toxicological Research, U.S. Food and Drug Administration, NCTR Drive, Jefferson, AR 72079

Ethel S. Gilbert, Pacific Northwest Laboratory, 902 Battelle Boulevard, Richland, WA 99352

Tony Godfrey, University of California San Francisco Mission Center Bldg., Room 234, 1855 Folsom St., San Francisco, CA 94103

John D. Graham, Harvard University, School of Public Health, Center for Risk Analysis, 677 Huntington Avenue, Boston, MA 02115

Peter G. Groer, University of Tennessee, Department of Nuclear Engineering, 210 Pasqua—Nuclear Engineering Building, Knoxville, TN 37996

Ronald W. Hart, National Center for Toxicological Research, Food and Drug Administration, NCTR Drive, Jefferson, AR 72079

John Higginson, Georgetown University Medical Center, Department of Community Medicine, 314 Kober-Cogan Hall, 3750 Reservoir Road, N.W., Washington, DC 20007

Colin K. Hill, University of Southern California School of Medicine, Department of Radiation Oncology and Microbiology, Albert Soiland Cancer Research Laboratory, CRL Building, Room 208B, 1303 N. Mission Street, Los Angeles, CA 90033

Harihara M. Mehendale, Division of Pharmacology and Toxicology, College of Pharmacy and Health Sciences, Northeast Louisiana University, Monroe, LA 71209-0470

George E. Milo, The Center for Molecular Environmental Health, Department of Medical Biochemistry and The Comprehensive Cancer Center, The Ohio State University, Columbus, OH 43210-1218

Myron Pollycove, United States Nuclear Regulatory Commission, OWFN 6H3, Washington, DC 20555

Leonard Sagan, Electric Power Research Institute, 3412 Hillview Avenue, Palo Alto, CA 94303

Kenneth F. Schaffner, The George Washington University, Philosophy

Department, 714 T Gelman Library, 2130 H Street, NW, Washington, DC 20052

Joan Smith-Sonneborn, Zoology & Physiology Department, Box 3166, University of Wyoming, Laramie, WY 82071

David J. Svendsgaard, Biostatistics Branch, Research Support Division (MD-55), Health Effects Research Laboratory, U.S. Environmental Protection Agency, Research Triangle Park, NC 27711

James E. Trosko, Michigan State University of Human Medicine, B-240 Life Sciences Building, East Lansing, MI 48824

Angelo Turturro, Division of Biometry and Risk Assessment, National Center for Toxicological Research, Food and Drug Administration, NCTR Drive, Jefferson, AR 72079

Index

ACDQ, *See* 6-Amino-7-chloro-5,8-dioxoquinoline
Acetaminophen, 126
Acetylaminofluorene, 207
Adaptive responses, 30, 255–266, 280–281
  atomic bomb survivors and, 255
  background radiation and, 257, 266
  DNA damage and repair and, 257, 266
  stress proteins and, 281
Aflatoxins, 56–57
Aging, 45, 248
AIG, *See* Anchorage independent growth
Alarmones, 247
Alcohol, 22, 101, 243, 244, 282
Altitude, 197
Ames, Bruce, 21, 22
6-Amino-7-chloro-5,8-dioxoquinoline (ACDQ), 247
Amphetamines, 80, *See also* specific types
Anchorage independent growth (AIG) stage, 43, 44, 45, 47, 52, 53, 57
Anomalies, 6
Anthralin, 49
Antibiotics, 31, 243–244, *See also* specific types
Anti-carcinogens, 218, *See also* specific types
Antimitosis, 131
Anti-oncogenes, 213, 218, *See also* specific types
Antioxidation, 93
Antiproliferative agents, 49, *See also* specific types
Anti-tumor promoters, 218, *See also* specific types
Apoptosis, 127, 209

Arndt-Schulz Law, 31, 38, 67, *See also* Nonmonotonic dose-response curves
Arsenic, 102
Arsenious acid, 31
Asbestos, 105, 207
Aspirin, 100–101, 282
Assays, *See also* specific types
  bio-, 90, 91, 92, 220, 274
  genotoxicity, 220
  mutation, 45–46
  NIH-3T3 transformation, 213
Atomic bomb survivors, 18, 155, 156–157, 164, 185, 282, *See also* Nuclear weapons fallout
  adaptive responses and, 255
  cancer in, 157, 173, 282
  leukemias in, 157, 173–176, 224, 282
  lifespan of, 176–177
  mortality in, 173–176, 177
  offspring of, 221
Atomic Energy Authority, 161
Atomic Weapons Establishment, 161
Autoimmune disease, 249, *See also* specific types
Autoprotection, 125–126
Avoidance reaction, 244

Background dose, 21, 88, 89, 93, 94, 171, 172, 208
  adaptive responses and, 257, 266
  spontaneous, 87
Backup safety systems, 17
Bacteria, 247
Bacterial endotoxins, 102, *See also* specific types
B(a)P, *See* Benzo(a)pyrene
Barker, Alan, 24

291

T - #0112 - 101024 - C0 - 234/156/17 [19] - CB - 9781566700931 - Gloss Lamination